미래 항공종사자들을 위한 항공기 정비 영어 입문서

항공기 정비 영어 기본편
Aircraft Maintenance English Basic

김형래

　항공정비영어는 항공기 정비뿐만 아니라 설계, 제작, 운항 등 항공산업의 여러 분야에서 필수적인 역량입니다. 이 책은 항공정비영어를 처음 접하는 입문자들이 정비 업무에 필요한 기본 영어를 체계적으로 학습할 수 있도록 구성되었습니다.

　초보 학습자들이 항공정비영어의 기본 구문을 이해하고, 주요 기술 용어를 정확히 연습하며, 항공기술에 대한 기초 지식을 쌓을 수 있도록 돕습니다. 특히 영어에 익숙하지 않은 독자들도 쉽게 따라갈 수 있도록 간결한 설명과 반복 학습 방식을 채택하여 부담 없이 학습할 수 있습니다.

　본 교재는 미국 연방항공청(FAA)의 Aviation Maintenance Technician Handbook 시리즈를 참고하여 항공기의 기본 개념을 쉽게 이해할 수 있도록 내용을 정리하였습니다. 주요 내용은 항공기 구조, 비행 조종면, 항공 역학, 복합 재료, 항공기 시스템, 엔진 및 프로펠러 등 항공기 정비에 필요한 기본적인 지식입니다.

　이 교재는 항공정비영어의 기초부터 차근차근 다져나갈 수 있도록 돕고, 이를 통해 실무에서 자신감을 가지고 영어를 효과적으로 활용할 수 있는 능력을 길러줍니다. 또한, 항공기 정비와 관련된 기본적인 기술 용어와 구문을 체계적으로 학습하여 현장에서도 바로 활용할 수 있도록 구성되었습니다.

　부족한 부분이나 보완이 필요한 내용에 대해 독자 여러분의 소중한 의견을 부탁드립니다. 이를 바탕으로 더욱 발전된 교재를 만들어 나가겠습니다.

　마지막으로, 이 책의 편찬을 위해 아낌없는 도움을 주신 연경문화사 이정수 대표님과 편집팀 여러분께 깊은 감사의 말씀을 드립니다.

저자 김형래

Features of the Book

01
실무 중심의 기초 영어 학습

항공정비 현장에서 실제로 사용되는 핵심 표현과
용어를 중심으로, 반복 학습을 통해 자연스럽게
구문을 이해하고 해석 능력을 키울 수 있습니다.

Words	adhesive 진착제 airframe 기체 aluminum 알루미늄 bolt 볼트 bulkhead 벌크헤드 (격벽) carbon fiber 탄소 섬유 composite 복 fabric 식물 fastener 제결 장치 fuselage 동체 impregnated fa landing gear 착륙 장치 longeron 롱거론 plywood 합판 rib 리 screw 나사 spar 스파 (주부재) stabilizer 안정판 stringer 스트 variety 다양성 welding 용접 wing 날개
Phrases	be constructed from ~로 제작되다 be joined by ~로 결합되다 consists of ~로 구성되다 ranging from ~ to ~ ~에서 ~까지 다양

02
**영어 초보자도
부담 없이 학습 가능**

영어에 익숙하지 않은 학습자도 부담 없이 접근할 수 있도록 간결한 설명과 친숙한 예문을 제공하여,
실무에 바로 적용할 수 있는 기초 지식을 확립할 수 있게 도와줍니다.

03
**음성 자료(MP3) 제공으로
듣기·말하기 연습**

영어 원문과 함께 제공되는 MP3 음성 파일을
활용하여 듣기와 말하기 능력을 향상시킬 수
있습니다. 소리 내어 읽으며 정확한 발음을 익히고,
유창성을 기를 수 있습니다.

1-1   General

di MP3 1-1

The airframe of a fixed-wing aircraft consists of five
wings, stabilizers, flight control surfaces, and landing g
Airframe structural components are constructed from
The earliest aircraft were constructed primarily of woo
common material, aluminum, followed. Many newly ce
molded composite materials, such as carbon fiber.

🔊 음원 서비스

본문의 모든 문장을 녹음한 음원을 제공합니다.
영어 문장을 2회 반복해서 들을 수 있습니다.
무한 반복하여 들으며 문장을 연습할 수 있습니다.

MP3 파일
QR코드에서
다운받을 수 있습니다.

유튜브
〈AeroTech English〉 채널에서
들을 수 있습니다.

04

학습 효과를 높이는 복습 문제 구성

각 장의 마지막에는 다양한 유형의 복습 문제를 수록하여
학습 내용을 점검하고 실전에 응용할 수 있는 기회를 제공합니다.
이를 통해 부족한 부분을 보완하며 학습 효과를 극대화할 수 있습니다.

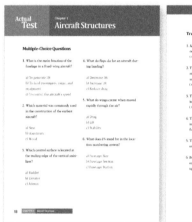

05

본문 전체 해석 제공으로
반복 학습 가능

본문 전체와 해석을 포함한 부록을 제공하여
언제든지 반복 학습이 가능하도록 구성하였습니다.
이를 통해 실무에서 필요한 실력을 쌓고,
학습한 내용을 깊이 이해할 수 있습니다.

Table of Contents

CHAPTER 1
Aircraft Structures

1-1　General

The airframe of a fixed-wing aircraft consists of five principal units: the fuselage, wings, stabilizers, flight control surfaces, and landing gear. [Figure 1-1]

Airframe structural components are constructed from a wide variety of materials. The earliest aircraft were constructed primarily of wood. Steel tubing and the most common material, aluminum, followed. Many newly certified aircraft are built from molded composite materials, such as carbon fiber.

Structural members of an aircraft's fuselage include stringers, longerons, ribs, bulkheads, and more. The main structural member in a wing is called the wing spar. The skin of aircraft can also be made from a variety of materials, ranging from impregnated fabric to plywood, aluminum, or composites.

The entire airframe and its components are joined by rivets, bolts, screws, and other fasteners. Welding, adhesives, and special bonding techniques are also used.

[Figure 1-1] Principal airframe units

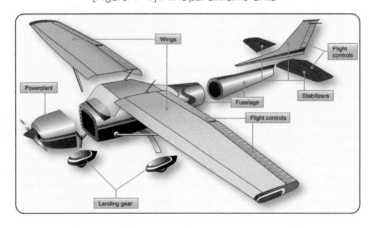

Words	adhesive 접착제　airframe 기체　aluminum 알루미늄　bolt 볼트 bulkhead 벌크헤드 (격벽)　carbon fiber 탄소 섬유　composite 복합 재료 fabric 직물　fastener 체결 장치　fuselage 동체　impregnated fabric 함침 직물 landing gear 착륙 장치　longeron 롱거론　plywood 합판　rib 리브　rivet 리벳 screw 나사　spar 스파 (주부재)　stabilizer 안정판　stringer 스트링거 variety 다양성　welding 용접　wing 날개
Phrases	be constructed from ～로 제작되다　be joined by ～로 결합되다 consists of ～로 구성되다　ranging from ~ to ~ ～에서 ～까지 다양하다

10　CHAPTER 1　Aircraft Structures

1-2 Fixed-Wing Aircraft

1) Fuselage

The fuselage is the main structure or body of the fixed-wing aircraft. It provides space for cargo, controls, accessories, passengers, and other equipment. In single-engine aircraft, the fuselage houses the powerplant. In multiengine aircraft, the engines may be either in the fuselage, attached to the fuselage, or suspended from the wing structure.

There are two general types of fuselage construction: truss and monocoque.

[Figure 1-2] A truss-type fuselage   [Figure 1-3] Monocoque construction

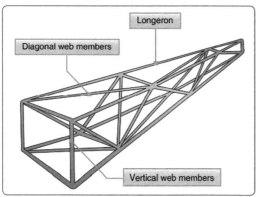 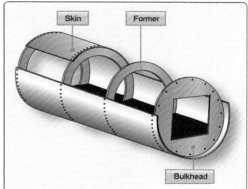

Words	accessories 부속품 cargo 화물 controls 조종 장치 engine 엔진 fixed-wing 고정익 fuselage 동체 monocoque 모노코크 multiengine 다발 엔진 passengers 승객 powerplant 동력 장치 single-engine 단발 엔진 structure 구조물 suspended 매달린 truss 트러스
Phrases	attached to ～에 부착된 may be either ～일 수 있다 provides space for ～을 위한 공간을 제공하다 suspended from ～에 매달린

2) Wing Configurations

Wings are airfoils that, when moved rapidly through the air, create lift. They are built in many shapes and sizes. Wing design can vary to provide certain desirable flight characteristics. Control at various operating speeds, the amount of lift generated, balance, and stability all change as the shape of the wing is altered.

The wing tip may be square, rounded, or even pointed. Figure 1-4 shows a number of typical wing leading and trailing edge shapes.

[Figure 1-4] Various wing design shapes

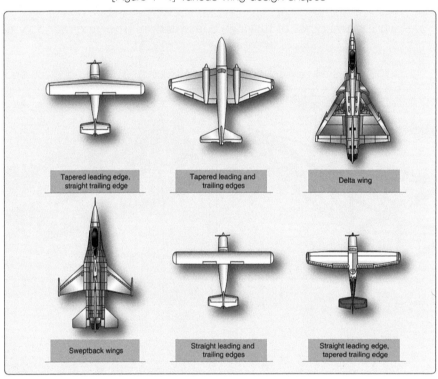

Words	airfoil 에어포일 altered 변경된 balance 균형 characteristic 특성 control 제어 create 생성하다 desirable 바람직한 edge 가장자리 generate 생성하다 leading edge 앞전 lift 양력 operating speed 작동 속도 pointed 뾰족한 rounded 둥근 stability 안정성 trailing edge 뒷전 vary 다양하다 wing tip 날개 끝
Phrases	as the shape of ~의 모양이 바뀔 때 can vary to provide ~을 제공하기 위해 다양하게 변경될 수 있다 shows a number of ~의 여러 가지를 보여준다

3) Empennage

The empennage of an aircraft is also known as the tail section. Most empennage designs consist of a tail cone, fixed aerodynamic surfaces or stabilizers, and movable aerodynamic surfaces.

The tail cone serves to close and streamline the aft end of most fuselages. The cone is made up of structural members like those of the fuselage; however, cones are usually of lighter construction since they receive less stress than the fuselage.

The other components of the typical empennage are of heavier construction than the tail cone. These members include fixed surfaces that help stabilize the aircraft and movable surfaces that help to direct an aircraft during flight. The fixed surfaces are the horizontal stabilizer and vertical stabilizer. The movable surfaces are usually a rudder located at the aft edge of the vertical stabilizer and an elevator located at the aft edge the horizontal stabilizer.

[Figure 1-5] The fuselage

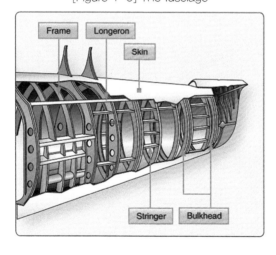

Words	aerodynamic 공기역학적 aft 후미 cone 콘 construction 제작 elevator 엘리베이터, 승강타 empennage 엠퍼너지, 미부, 미익 fuselage 동체 horizontal stabilizer 수평 안정판 member 구성 요소 movable 가동식 rudder 러더, 방향타 stabilizer 안정판 streamline 유선형으로 만들다 structure 구조 tail cone 꼬리 콘 vertical stabilizer 수직 안정판
Phrases	be made up of ~로 구성되다 be usually of 보통 ~로 되어 있다 located at ~에 위치하다 serves to ~의 역할을 하다

4) Location Numbering Systems

The applicable manufacturer's numbering system and abbreviated designations or symbols should always be reviewed before attempting to locate a structural member. They are not always the same. The following list includes location designations typical of those used by many manufacturers.

• Fuselage stations (Fus. Sta. or FS) are numbered in inches from a reference or zero point known as the reference datum.

• Buttock line or butt line (BL) is a vertical reference plane down the center of the aircraft from which measurements left or right can be made.

• Water line (WL) is the measurement of height in inches perpendicular from a horizontal plane usually located at the ground, cabin floor, or some other easily referenced location.

[Figure 1-6] The various fuselage stations

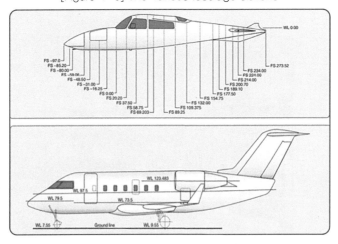

Words	**abbreviated** 단축된, 생략된 **applicable** 적용 가능한 **buttock line (BL)** 버턱 라인(버트 라인) **cabin floor** 객실 바닥 **datum** 기준점 **designation** 명칭 **fuselage station (FS)** 동체 기준점 **horizontal plane** 수평면 **manufacturer** 제조사 **measurement** 측정 **perpendicular** 수직의 **reference** 기준 **reference datum** 기준 데이터(기준점) **structural member** 구조 부재 **symbol** 기호 **vertical reference plane** 수직 기준 평면 **water line (WL)** 워터 라인
Phrases	**be numbered in** ~로 번호가 매겨지다 **from which measurements can be made** ~에서 측정이 가능하다 **is the measurement of** ~의 측정값이다 **known as** ~로 알려진

1-3 Flight Control Surfaces

1) Description

The directional control of a fixed-wing aircraft takes place around the lateral, longitudinal, and vertical axes by means of flight control surfaces designed to create movement about these axes. These control devices are hinged or movable surfaces through which the attitude of an aircraft is controlled during takeoff, flight, and landing. They are usually divided into two major groups: primary or main flight control surfaces and secondary or auxiliary control surfaces.

The primary flight control surfaces on a fixed-wing aircraft include: ailerons, elevators, and the rudder. The ailerons are attached to the trailing edge of both wings and when moved, rotate the aircraft around the longitudinal axis. The elevator is attached to the trailing edge of the horizontal stabilizer. When it is moved, it alters aircraft pitch, which is the attitude about the horizontal or lateral axis. The rudder

[Figure 1–7] Flight control surfaces move the aircraft around the three axes of flight

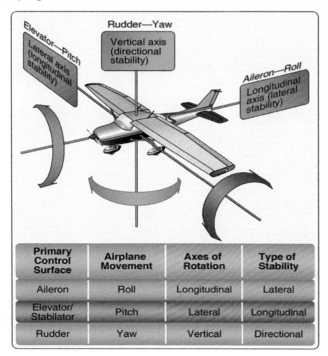

Primary Control Surface	Airplane Movement	Axes of Rotation	Type of Stability
Aileron	Roll	Longitudinal	Lateral
Elevator/ Stabilator	Pitch	Lateral	Longitudinal
Rudder	Yaw	Vertical	Directional

is hinged to the trailing edge of the vertical stabilizer. When the rudder changes position, the aircraft rotates about the vertical axis (yaw). Figure 1-7 shows the primary flight controls of a light aircraft and the movement they create relative to the three axes of flight.

Words	aileron 에일러론 **auxiliary** 보조의 **axis** 축 **hinged** 힌지로 연결된 **longitudinal axis** 종축 **lateral axis** 횡축 **movement** 움직임 **pitch** 피치(상하 자세) **primary** 주요 **trailing edge** 뒷전 **vertical axis** 수직축 **vertical stabilizer** 수직 안정판 **yaw** 요(좌우 회전)
Phrases	**be attached to** ~에 부착되어 있다 **be hinged to** ~에 힌지로 연결되어 있다 **by means of** ~수단으로 **divided into** ~로 나뉘다 **relative to** ~에 따라 **take place** 이루어지다 **through which** 이를 통해

2) Flaps

Flaps are found on most aircraft. They are usually inboard on the wings' trailing edges adjacent to the fuselage. Leading edge flaps are also common. They extend forward and down from the inboard wing leading edge. The flaps are lowered to increase the camber of the wings and provide greater lift and control at slow speeds. They enable landing at slower speeds and shorten the amount of runway required for takeoff and landing.

Flaps are usually constructed of materials and with techniques used on the other airfoils and control surfaces of a particular aircraft. Aluminum skin and structure flaps are the norm on light aircraft. Heavy and high-performance aircraft flaps may also be aluminum, but the use of composite structures is also common.

[Figure 1-8] Various aircraft and type with flaps

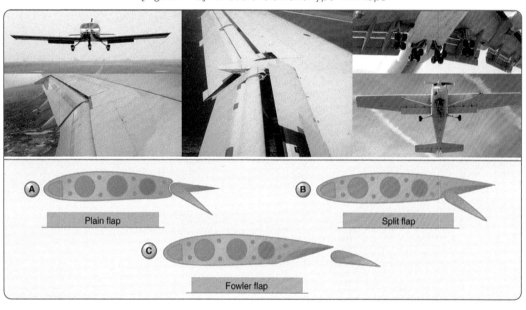

Words	airfoil 공기역학적 표면 aluminum 알루미늄 camber 캠버 (윙의 곡률) composite 복합 재료 control surface 제어 표면 edge 모서리 extend 확장하다 fuselage 동체 flap 플랩 leading edge 앞전(선두 모서리) lift 양력 materials 재료 runway 활주로 structure 구조 trailing edge 뒷전(후방 모서리)
Phrases	adjacent to ～에 인접한 at slow speeds 저속에서 be found on ～에 있다 be lowered 내려가다 be used on ～에 사용되다 extend forward and down 앞으로 내려가다 inboard on ～의 내부에

Actual
Test

Chapter 1

Aircraft Structures

Multiple-Choice Questions

1. What is the main function of the fuselage in a fixed-wing aircraft?

   a) To generate lift
   b) To hold passengers, cargo, and equipment
   c) To control the aircraft's speed

2. Which material was commonly used in the construction of the earliest aircraft?

   a) Steel
   b) Aluminum
   c) Wood

3. Which control surface is located at the trailing edge of the vertical stabilizer?

   a) Rudder
   b) Elevator
   c) Aileron

4. What do flaps do for an aircraft during landing?

   a) Decrease lift
   b) Increase lift
   c) Reduce drag

5. What do wings create when moved rapidly through the air?

   a) Drag
   b) Lift
   c) Stability

6. What does FS stand for in the location numbering system?

   a) Fuselage Size
   b) Fuselage Section
   c) Fuselage Station

True/False (O/X) Questions

1. All aircraft are constructed using only aluminum and steel materials. (O / X)

2. The tail cone of an aircraft is usually made of heavy materials because it supports the weight of the engines. (O / X)

3. The elevator is attached to the trailing edge of the horizontal stabilizer. (O / X)

4. The primary flight control surfaces include the ailerons, rudder, and flaps. (O / X)

5. The wing tip can be square, rounded, or pointed. (O / X)

6. Buttock line (BL) is a horizontal reference plane from which measurements up or down can be made. (O / X)

Short Answer Questions

1. Which component helps stabilize the aircraft's pitch?

2. What is the purpose of the ailerons on an aircraft?

3. What is the material commonly used in the construction of light aircraft wings?

4. Which surface is responsible for controlling aircraft yaw?

5. What do wings generate for an aircraft to fly?

6. Where is the water line usually referenced from?

Multiple-Choice Questions

1. b) To hold passengers, cargo, and equipment
2. c) Wood
3. a) Rudder
4. b) Increase lift and control at slow speeds
5. b) Lift
6. c) Fuselage Station

True/False (O/X) Questions

1. X (Aircraft are made from various materials, including wood, aluminum, composites, and steel.)
2. X (The tail cone is usually of lighter construction because it receives less stress than the fuselage.)
3. O
4. X (The primary flight control surfaces are the ailerons, rudder, and elevator, not flaps.)
5. O
6. X (Explanation: Buttock line (BL) is a vertical reference plane, not a horizontal one.)

Short Answer Questions

1. Horizontal stabilizer
2. To control the aircraft's roll
3. Aluminum
4. Rudder
5. Lift
6. Ground, cabin floor, or another easily referenced location

CHAPTER 2
Aerodynamics

1) Bernoulli's Principle

Bernoulli's principle states that when a fluid (air) flowing through a tube reaches a constriction, or narrowing, of the tube, the speed of the fluid flowing through that constriction is increased and its pressure is decreased. The cambered (curved) surface of an airfoil (wing) affects the airflow exactly as a constriction in a tube affects airflow.

Diagram A of Figure 2-1 illustrates the effect of air passing through a constriction in a tube. In Diagram B, air is flowing past a cambered surface, such as an airfoil, and the effect is similar to that of air passing through a restriction.

As the air flows over the upper surface of an airfoil, its velocity increases and its pressure decreases; an area of low pressure is formed. There is an area of greater pressure on the lower surface of the airfoil, and this greater pressure tends to move the wing upward. The difference in pressure between the upper and lower surfaces of the wing is called lift.

[Figure 2-1] Bernoulli's Principle

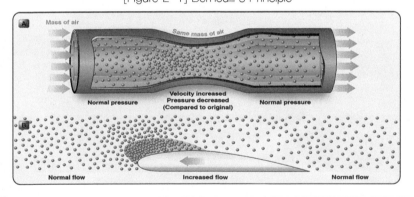

Words	**airfoil** 날개 **airflow** 공기 흐름 **cambered** 곡면의 **constriction** 좁아지는 부분 **decrease** 감소하다 **diagram** 도면 **fluid** 유체 **increase** 증가하다 **lift** 양력 **pressure** 압력 **restriction** 제한, 좁은 부분 **velocity** 속도
Phrases	**flow through** ~을 통과하다 **such as** ~와 같은 **tends to** ~하는 경향이 있다 **the difference in pressure** 압력 차이

2) Airfoil

An airfoil is any device that creates a force, based on Bernoulli's principles or Newton's laws, when air is caused to flow over the surface of the device. An airfoil can be the wing of an airplane, the blade of a propeller, the rotor blade of a helicopter, or the fan blade of a turbofan engine. The wing of an airplane moves through the air because the airplane is in motion, and generates lift by the process previously described.

In Figure 2-2 an airfoil, or wing, is shown, with some of the terminology that is used to describe a wing.

[Figure 2–2] Wing terminology

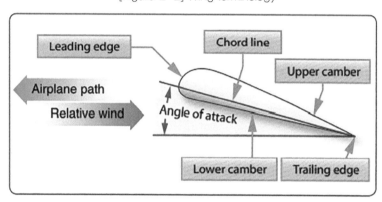

Words	**airfoil** 에어포일 **Bernoulli's principle** 베르누이의 원리 **blade** 날 **fan blade** 팬 날개 **helicopter** 헬리콥터 **Newton's laws** 뉴턴의 법칙 **propeller** 프로펠러 **rotor blade** 로터 날개 **surface** 표면 **terminology** 용어 **turbofan engine** 터보팬 엔진
Phrases	**based on** ~을 기반으로 **caused to flow** 흐르게 하다 **in motion** 움직이는 **moves through** 지나가다

2-2 Forces in action during flight

◁)) MP3 2-2

In all types of flying, flight calculations are based on the magnitude and direction of four forces: weight, lift, drag, and thrust.

An aircraft in flight is acted upon by four forces.

[Figure 2-3] Forces in action during flight

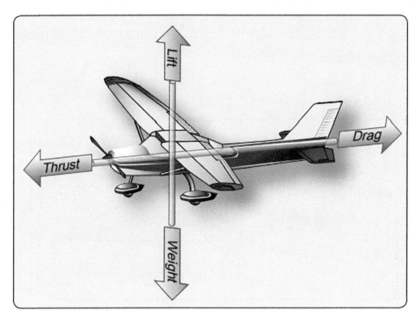

1) Gravity or weight

the force that pulls the aircraft toward the earth. Weight is the force of gravity acting downward upon everything that goes into the aircraft, such as the aircraft itself, crew, fuel, and cargo.

2) Lift

the force that pushes the aircraft upward. Lift acts vertically and counteracts the effects of weight.

3) Thrust

the force that moves the aircraft forward. Thrust is the forward force produced by the powerplant that overcomes the force of drag.

4) Drag

the force that exerts a braking action to hold the aircraft back. Drag is a backward deterrent force and is caused by the disruption of the airflow by the wings, fuselage, and protruding objects.

Words	**aircraft** 항공기 **cargo** 화물 **counteracts** 상쇄하다 **deterrent** 억제력, 방해 **drag** 항력 **forces** 힘 **forward** 전진 **gravity** 중력 **overcome** 이겨내다 **powerplant** 동력장치(파워플랜트) **thrust** 추진력 **weight** 무게
Phrases	**act upon** ~에 영향을 미치다 **be based on** ~을 바탕으로 하다 **break action** 제동 작용 **exert a force** 힘을 가하다 **hold back** 뒤로 제지하다 **pull toward** ~로 끌어당기다 **push upward** 위로 밀다

The Axes of an Aircraft

Whenever an aircraft changes its attitude in flight, it must turn about one or more of three axes. Figure 2-4 shows the three axes, which are imaginary lines passing through the center of the aircraft.

The axes of an aircraft can be considered as imaginary axles around which the aircraft turns like a wheel. At the center, where all three axes intersect, each is perpendicular to the other two. The axis that extends lengthwise through the fuselage from the nose to the tail is called the longitudinal axis. The axis that extends crosswise from wing tip to wing tip is the lateral, or pitch, axis. The axis that passes through the center, from top to bottom, is called the vertical, or yaw, axis. Roll, pitch, and yaw are controlled by three control surfaces.

Roll is produced by the ailerons, which are located at the trailing edges of the wings. Pitch is affected by the elevators, the rear portion of the horizontal tail assembly. Yaw is controlled by the rudder, the rear portion of the vertical tail assembly.

Words	**aileron** 에일러론 **axis** 축 **crosswise** 가로로 **elevator** 엘리베이터 **extend** 연장하다, 뻗어 있다 **fuselage** 동체 **imaginary** 가상의 **lateral axis** 측면 축, 가로축 **lengthwise** 세로로, 길게 **longitudinal axis** 종축, 세로축 **pitch** 피치 **rudder** 러더 **roll** 롤 **tail assembly** 꼬리날개 조립체 **vertical axis** 수직축 **wing tip** 날개 끝 **yaw** 요
Phrases	**about one or more of** 하나 또는 그 이상의 **controlled by** ~에 의해 제어되다 **is called** ~라고 불린다 **located at** ~에 위치하다 **produced by** ~에 의해 발생하다 **the rear portion of** ~의 후방 부분

[Figure 2-4] Motion of an aircraft about its axes

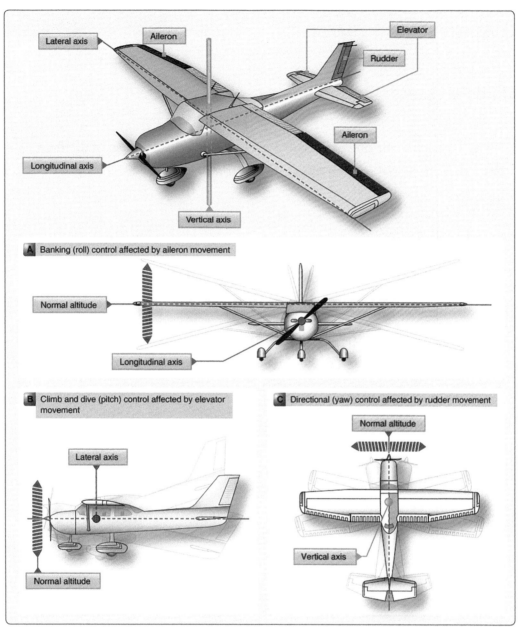

A Banking (roll) control affected by aileron movement

B Climb and dive (pitch) control affected by elevator movement

C Directional (yaw) control affected by rudder movement

Multiple-Choice Questions

1. What happens to air pressure as air flows over the upper surface of an airfoil?

   a) It increases

   b) It decreases

   c) It stays the same

2. Which principle explains the creation of lift on an airfoil?

   a) Newton's third law

   b) Bernoulli's principle

   c) Conservation of momentum

3. What force counteracts weight during flight?

   a) Lift

   b) Drag

   c) Thrust

4. What is the purpose of the drag force?

   a) To push the aircraft upward

   b) To move the aircraft forward

   c) To hold the aircraft back

5. What force allows the aircraft to move forward?

   a) Thrust

   b) Lift

   c) Drag

6. How many forces act on an aircraft in flight?

   a) Two

   b) Four

   c) Six

True/False (O/X) Questions

1. Lift is the force that counteracts weight and acts vertically. (O / X)

2. Thrust is generated by the wings of the aircraft. (O / X)

3. The rudder controls the roll of an aircraft. (O / X)

4. Elevators affect the pitch of an aircraft. (O / X)

5. The lateral axis runs from the nose to the tail of the aircraft. (O / X)

6. The vertical axis controls the yaw motion of an aircraft. (O / X)

Short Answer Questions

1. What principle explains the effect of air pressure and velocity on an airfoil?

2. Which control surface affects roll?

3. What axis runs from top to bottom through the aircraft?

4. Which axis controls the pitch of an aircraft?

5. What is the difference in pressure between the upper and lower surfaces of a wing called?

6. Where is the center of gravity on an aircraft?

Multiple-Choice Questions

1. b) It decreases
2. b) Bernoulli's principle
3. a) Lift
4. c) To hold the aircraft back
5. a) Thrust
6. b) Four

True/False (O/X) Questions

1. O
2. X (Thrust is generated by the powerplant, not the wings.)
3. X (The rudder controls the yaw of an aircraft, not roll.)
4. O
5. X (The lateral axis runs crosswise from wing tip to wing tip, not nose to tail.)
6. O

Short Answer Questions

1. Bernoulli's principle
2. Ailerons
3. Vertical axis
4. Lateral axis
5. Lift
6. At the center where all axes intersect

CHAPTER 3
Advanced Composite Materials

3-1 General

Composite materials are becoming more important in the construction of aerospace structures. Aircraft parts made from composite materials, such as fairings, spoilers, and flight controls, were developed during the 1960s for their weight savings over aluminum parts.

New generation large aircraft are designed with all composite fuselage and wing structures, and the repair of these advanced composite materials requires an in-depth knowledge of composite structures, materials, and tooling. The primary advantages of composite materials are their high strength, relatively low weight, and corrosion resistance.

[Figure 3-1] A380 aircraft composite applications

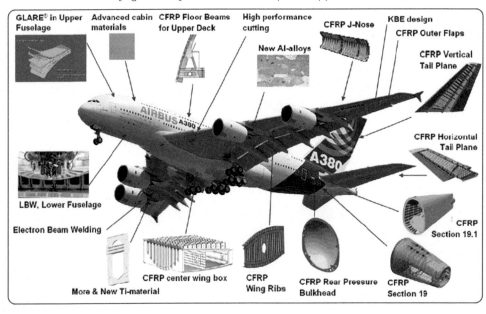

Words	aerospace 항공 우주 corrosion resistance 내식성 composite 복합 재료 fairings 페어링 flight controls 비행 조종 장치 materials 재료 relatively 상대적으로 repair 수리 strength 강도 spoilers 스포일러 tooling 치공구, 도구 weight savings 무게 절감 wing structures 날개 구조
Phrases	developed during ~동안 개발되었다 primary advantages of ~의 주요 장점 requires an in-depth knowledge of ~에 대한 심층적인 지식이 필요하다 weight savings over ~에 비해 무게 절감

3-2 Composite Material and Process

◁» MP3 3-2

1) Prepreg

Prepreg is a fabric or tape that is impregnated with a resin during the manufacturing process. The resin system is already mixed and is in the B stage cure. Store the prepreg material in a freezer below 0 °F to prevent further curing of the resin.

You must remove the prepreg from the freezer and let the material thaw, which might take 8 hours for a full roll. Store the prepreg materials in a sealed, moisture proof bag. Do not open these bags until the material is completely thawed, to prevent contamination of the material by moisture.

[Figure 3-2] Freezer for storing prepreg materials

Words	above ～위에 bag 가방, 백 below ～아래에 curing 경화 fabric 직물 freezer 냉동고 moisture 수분 prepreg 프리프레그 resin 수지, 레진 sealed 밀봉된 tape 테이프
Phrases	be impregnated with ～로 침투된 in the B stage cure B단계 경화 상태 store in ～에 보관하다 take ~ hours ～시간이 걸리다 to prevent ~ ～을 방지하기 위해

2) Vacuum Bag

The vacuum bag material provides a tough layer between the repair and the atmosphere. The vacuum bag material is available in different temperature ratings, so make sure that the material used for the repair can handle the cure temperature. Most vacuum bag materials are one time use, but material made from flexible silicon rubber is reusable.

[Figure 3–3] Bagging of complex part

Words	atmosphere 대기 bag 백 cure 경화 flexible 유연한 reusable 재사용 가능 rubber 고무 silicon 실리콘 temperature 온도 tough 강한
Phrases	handle the cure temperature 경화 온도를 처리하다 one time use 한 번만 사용하다 temperature ratings 온도 등급 vacuum bag 진공 백

3-3　Composite Cure Equipment

1) Autoclave

An autoclave system allows a complex chemical reaction to occur inside a pressure vessel according to a specified time, temperature, and pressure profile in order to process a variety of materials.

[Figure 3-4] Autoclave

Autoclaves that are operated at lower temperatures and pressures can be pressurized by air, but if higher temperatures and pressures are required for the cure cycle, a 50/50 mixture of air and nitrogen or 100 percent nitrogen should be used to reduce the change of an autoclave fire.

The major elements of an autoclave system are a vessel to contain pressure, sources to heat the gas stream and circulate it uniformly within the vessel, a subsystem to apply vacuum to parts covered by a vacuum bag, a subsystem to control operating parameters, and a subsystem to load the molds into the autoclave.

Modern autoclaves are computer controlled and the operator can write and monitor all types of cure cycle programs. The most accurate way to control the cure cycle is to control the autoclave controller with thermocouples that are placed on the actual part.

Words	**air** 공기　**autoclave** 오토클레이브　**chemical reaction** 화학 반응　**control** 제어　**cure cycle** 경화 사이클　**gas stream** 가스 흐름　**mold** 몰드　**nitrogen** 질소　**pressure vessel** 압력 용기　**subsystem** 서브시스템　**thermocouple** 열전대　**vacuum** 진공
Phrases	**a mixture of** ~의 혼합물　**be operated at** ~에서 작동되다　**set the parameters** 매개변수를 설정하다　**uniform circulation** 균일한 순환　**write and monitor** 작성하고 모니터링하다

2) Heat Bonder

A heat bonder is a portable device that automatically controls heating based on temperature feedback from the repair area. Heat bonders also have a vacuum pump that supplies and monitors the vacuum in the vacuum bag.

[Figure 3–5] Heat bonder equipment

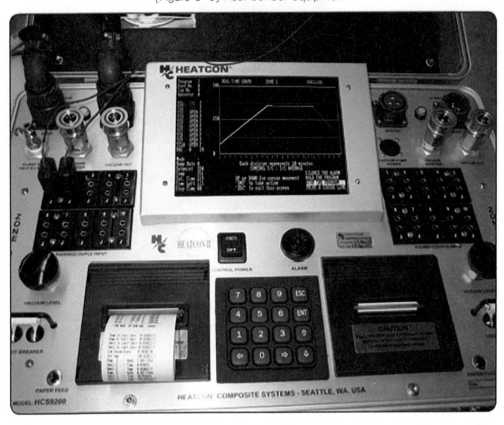

Heat Blanket

A heat blanket is a flexible heater. It is made of two layers of silicon rubber with a metal resistance heater between the two layers of silicon. Heat blankets are a common method of applying heat for repairs on the aircraft.

[Figure 3-6] Heat blankets

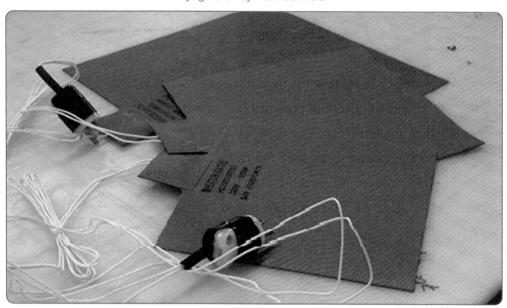

Words	**blanket** 블랭킷 **device** 장치 **flexible** 유연한 **heater** 히터 **heat** 열 **metal** 금속 **pump** 펌프 **repair** 수리 **rubber** 고무 **silicon** 실리콘 **temperature** 온도 **vacuum pump** 진공 펌프
Phrases	**a common method of** ~의 일반적인 방법 **based on** ~에 기반하여 **flexible heater** 유연한 히터 **monitors the vacuum** 진공을 모니터링하다 **supplies the vacuum** 진공을 공급하다

1) Ultrasonic Bond Tester Inspection

Low-frequency and high-frequency bond testers are used for ultrasonic inspections of composite structures. These bond testers use an inspection probe that has one or two transducers.

The high-frequency bond tester is used to detect delaminations and voids. It cannot detect a skin-to-honeycomb core disbond or porosity. It can detect defects as small as 0.5-inch in diameter.

The low-frequency bond tester uses two transducers and is used to detect delamination, voids, and skin to honeycomb core disbands. This inspection method does not detect which side of the part is damaged, and cannot detect defects smaller than 1.0-inch.

[Figure 3-7] Bond tester

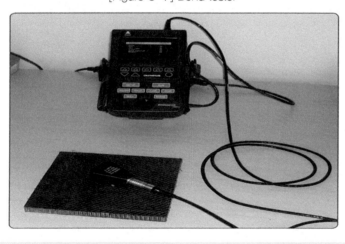

Words	**bond tester** 본드 시험기, 결함 시험기 **core** 코어 **delamination** 박리 **defect** 결함 **detect** 검출하다 **diameter** 지름 **disband** 해체하다 **frequency** 주파수 **honeycomb** 벌집 **inspection** 검사 **porosity** 기공 **probe** 탐침 **transducer** 변환기 **void** 공간, 공극
Phrases	**as small as** ~만큼 작은 **be used for** ~를 위해 사용되다 **be used to** ~하는 데 사용되다

2) Audible Sonic Testing (Coin Tapping)

Sometimes referred to as audio, sonic, or coin tap, this technique makes use of frequencies in the audible range (10 Hz to 20 Hz). A surprisingly accurate method in the hands of experienced personnel, tap testing is perhaps the most common technique used for the detection of delamination and/or disbond.

[Figure 3–8] Tap test with tap hammer

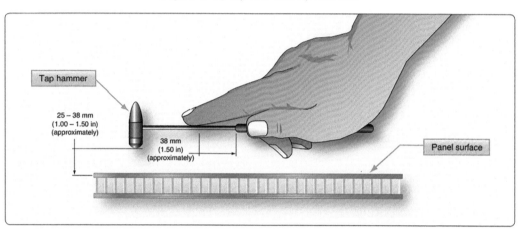

Words	**accurate** 정확한 **audible** 가청의 **delamination** 박리 **disbond** 탈착하다, 분리하다 **frequencies** 주파수 **personnel** 인력 **sonic** 소리의, 음파의 **technique** 기술 **tap testing** 동전 두드리기 테스트
Phrases	**in the hands of** ～의 손에 달려 있다 **make use of** ～을 사용하다 **referred to as** ～라고 불리다

Multiple-Choice Questions

1. What is the primary advantage of composite materials in aerospace structures?

   a) High cost
   b) Short lifespan
   c) Corrosion resistance

2. Where should prepreg materials be stored to prevent curing?

   a) At room temperature
   b) In a freezer below 0°F
   c) In direct sunlight

3. What is the main function of a vacuum bag?

   a) To heat composite materials
   b) To store prepreg materials
   c) To protect the repair from the atmosphere

4. What gas is recommended for high-temperature autoclave operations?

   a) Oxygen
   b) Nitrogen
   c) Carbon dioxide

5. What does the low-frequency bond tester detect?

   a) Skin-to-honeycomb core disbonds
   b) Defects smaller than 0.5-inch
   c) Only surface scratches

6. What does a coin-tap test rely on?

   a) Temperature feedback
   b) Audible sound differences
   c) High-frequency ultrasonic waves

True/False (O/X) Questions

1. Composite materials are heavier than aluminum parts. (O / X)

2. Vacuum bags are reusable in all cases. (O / X)

3. Heat blankets are a common method for applying heat during repairs. (O / X)

4. Coin tapping is effective for detecting delamination. (O / X)

5. The low-frequency bond tester can detect defects smaller than 0.5-inch. (O / X)

6. Prepreg materials should be stored in sealed, moisture-proof bags. (O / X)

Short Answer Questions

1. What is the main advantage of composite materials in aerospace structures?

2. Why should you let prepreg materials thaw before opening the bag?

3. What is the primary use of a vacuum bag in repairs?

4. What is an autoclave used for?

5. What does coin tapping help identify?

6. What is inside a heat blanket for heating?

Multiple-Choice Questions

1. c) Corrosion resistance
2. b) In a freezer below 0°F
3. c) To protect the repair from the atmosphere
4. b) Nitrogen
5. a) Skin-to-honeycomb core disbonds
6. b) Audible sound differences

True/False (O/X) Questions

1. X (Composite materials are lighter than aluminum parts.)
2. X (Most vacuum bag materials are one-time use, but some silicon rubber types are reusable.)
3. O
4. O
5. X (The low-frequency bond tester cannot detect defects smaller than 1.0-inch.)
6. O

Short Answer Questions

1. High strength, low weight, and corrosion resistance.
2. To prevent moisture contamination.
3. To provide a protective layer between the repair and the atmosphere.
4. For curing composite materials under controlled pressure, temperature, and time.
5. Delamination or disbonding in composite materials.
6. A resistance heater between silicon rubber layers.

CHAPTER 4
Aircraft Systems

Aircraft Instrument Systems

1) Introduction

Since the beginning of manned flight, it has been recognized that supplying the pilot with information about the aircraft and its operation could be useful and lead to safer flight. The Wright Brothers had very few instruments on their Wright Flyer, but they did have an engine tachometer, an anemometer (wind meter), and a stop watch. They were obviously concerned about the aircraft's engine and the progress of their flight.

From that simple beginning, a wide variety of instruments have been developed to inform flight crews of different parameters. Instrument systems now exist to provide information on the condition of the aircraft, engine, components, the aircraft's attitude in the sky, weather, cabin environment, navigation, and communication.

Words	**anemometer** 풍속계 (바람 측정기) **attitude** 자세 **cabin environment** 객실 환경 **components** 구성 요소 **engine** 엔진 **instrument** 계기 **navigation** 항법 **parameters** 매개변수 **stop watch** 스톱워치 **tachometer** 회전계 (엔진 회전수 측정기) **weather** 날씨
Phrases	**at the beginning** 처음에 **concerned about** ~에 대해 걱정하다 **lead to** ~로 이어지다 **provide information on** ~에 대한 정보를 제공하다

[Figure 4–1] Wright Flyer instruments, Boeing 707 airliner cockpit,
and an Airbus A380 glass cockpit

2) Classifying Instruments

There are three basic kinds of instruments classified by the job they perform: flight instruments, engine instruments, and navigation instruments. There are also miscellaneous gauges and indicators that provide information that do not fall into these classifications, especially on large complex aircraft. Flight control position, cabin environmental systems, electrical power, and auxiliary power units (APUs), for example, are all monitored and controlled from the cockpit via the use of instruments systems.

Words	**auxiliary power units (APUs)** 보조 동력 장치 **cockpit** 조종석 **electrical power** 전기력 **engine instruments** 엔진 계기 **flight instruments** 비행 계기 **flight control position** 비행 제어 위치 **gauges** 게이지 **indicators** 지시기, 표시기 **navigation instruments** 항법 계기
Phrases	**classified by** ~에 따라 분류된 **fall into** ~에 속하다 **for example** 예를 들어 **via the use of** ~의 사용을 통해

(1) Flight Instruments

The instruments used in controlling the aircraft's flight attitude are known as the flight instruments. There are basic flight instruments, such as the altimeter that displays aircraft altitude; the airspeed indicator; and the magnetic direction indicator, a form of compass. Additionally, an artificial horizon, turn coordinator, and vertical speed indicator are flight instruments present in most aircraft.

Over the years, flight instruments have come to be situated similarly on the instrument panels in most aircraft. This basic T arrangement for flight instruments is shown in Figure 4-2. The top center position directly in front of the pilot and copilot is the basic display position for the artificial horizon even in modern glass cockpits (those with solid-state, flat-panel screen indicating systems).

[Figure 4-2] The basic T arrangement of analog flight instruments

Words	**airspeed indicator** 속도계　**artificial horizon** 인공 수평선 계기　**altitude** 고도 **compass** 나침반　**copilot** 부조종사　**flight instruments** 비행 계기 **magnetic direction indicator** 자석 방향 표시기　**pilot** 조종사 **turn coordinator** 선회계(회전 조정기)　**vertical speed indicator** 수직 속도 표시기
Phrases	**be known as** ~로 알려져 있다　**even in** ~에서도　**come to be** ~하게 되다 **directly in front of** ~의 바로 앞에　**shown in** ~에서 보여지다 **situated similarly** 비슷하게 배치되다

(2) Engine Instruments

Engine instruments are those designed to measure operating parameters of the aircraft's engine. These are usually quantity, pressure, and temperature indications. They also include measuring engine speed. The most common engine instruments are the fuel and oil quantity and pressure gauges, tachometers, and temperature gauges.

Engine instrumentation is often displayed in the center of the cockpit where it is easily visible to the pilot and copilot. On light aircraft requiring only one flight crew member, this may not be the case. Multiengine aircraft often use a single gauge for a particular engine parameter, but it displays information for all engines through the use of multiple pointers on the same dial face.

[Figure 4-3] An engine instrumentation

Words	engine 엔진 gauges 게이지 oil 오일 pressure 압력 quantity 양 tachometer 타코미터(엔진 회전수 계기) temperature 온도 visible 볼 수 있는
Phrases	be designed to ~하도록 설계되다 be displayed in ~에 표시되다 flight crew 비행 승무원 in the center of ~의 중앙에 through the use of ~의 사용을 통해

(3) Navigation Instruments

Navigation instruments are those that contribute information used by the pilot to guide the aircraft along a definite course. This group includes compasses of various kinds, some of which incorporate the use of radio signals to define a specific course while flying the aircraft en route from one airport to another. Other navigational instruments are designed specifically to direct the pilot's approach to landing at an airport. Traditional navigation instruments include a clock and a magnetic compass.

Along with the airspeed indicator and wind information, these can be used to calculate navigational progress. Radios and instruments sending locating information via radio waves have replaced these manual efforts in modern aircraft.

Global position systems (GPS) use satellites to pinpoint the location of the aircraft via geometric triangulation. This technology is built into some aircraft instrument packages for navigational purposes.

[Figure 4–4] Navigation instruments

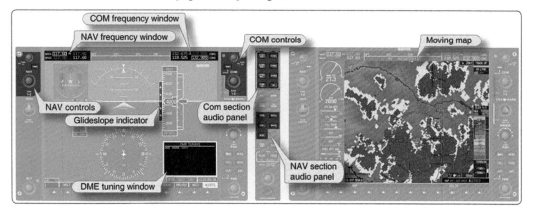

Words	aircraft 항공기 compass 나침반 global position system 글로벌 위치 시스템 (GPS) instrument 계기 location 위치 magnetic compass 자석 나침반 radio waves 라디오 파장 satellites 위성 triangulation 삼각측량
Phrases	along with ~와 함께 en route ~로 가는 중(도중에) in modern aircraft 현대 항공기에서 pinpoint the location 위치를 정확하게 찾다

Aircraft Instrument Systems

Multiple-Choice Questions

1. What is the primary purpose of flight instruments?

   a) To measure engine temperature
   b) To control the aircraft's flight attitude
   c) To guide the aircraft along a definite course

2. What do engine instruments typically measure?

   a) Aircraft altitude
   b) Weather conditions
   c) Engine parameters like quantity, pressure, and temperature

3. What is the role of navigation instruments?

   a) To measure engine pressure
   b) To control cabin environmental systems
   c) To provide information for guiding the aircraft along a course

4. What technology does GPS use for navigation?

   a) Magnetic fields
   b) Geometric triangulation with satellites
   c) Engine performance signals

5. What does a turn coordinator help the pilot to control?

   a) Aircraft altitude
   b) Fuel pressure
   c) Aircraft turning and bank attitude

6. Which flight instrument shows aircraft altitude?

   a) Altimeter
   b) Tachometer
   c) Artificial horizon

True/False (O/X) Questions

1. The Wright Brothers used an altimeter in their first flight. (O / X)

2. Engine instruments are used to guide the aircraft along a definite course. (O / X)

3. The artificial horizon is typically located in the top center position of the instrument panel. (O / X)

4. Navigation instruments can use radio signals to define a course. (O / X)

5. The main job of navigation instruments is to measure fuel levels. (O / X)

6. Traditional navigation instruments include a clock and magnetic compass. (O / X)

Short Answer Questions

1. What device measures aircraft altitude?

2. What do engine instruments typically display?

3. What does a tachometer measure?

4. What is one purpose of navigation instruments?

5. Name one example of a traditional navigation instrument.

6. Name one instrument used by the Wright Brothers during their first flight.

Multiple-Choice Questions

1. b) To control the aircraft's flight attitude

2. c) Engine parameters like quantity, pressure, and temperature

3. c) To provide information for guiding the aircraft along a course

4. b) Geometric triangulation with satellites

5. c) Aircraft turning and bank attitude

6. a) Altimeter

True/False (O/X) Questions

1. X (They used an anemometer, not an altimeter.)

2. X (Engine instruments measure engine parameters, not navigation.)

3. O

4. O

5. X (Navigation instruments guide the aircraft, not measure fuel levels.)

6. O

Short Answer Questions

1. Altimeter

2. Quantity, pressure, temperature, or engine speed

3. Engine speed

4. To guide the aircraft along a definite course

5. Clock or magnetic compass

6. Anemometer (or tachometer, stop watch)

4-2 Aircraft Hydraulic Systems

◁》 MP3 4-2-1

1) Introduction

Hydraulic systems are not new to aviation. Early aircraft had hydraulic brake systems. Hydraulic systems in aircraft provide a means for the operation of aircraft components. The operation of landing gear, flaps, flight control surfaces, and brakes is largely accomplished with hydraulic power systems. Hydraulic system complexity varies from small aircraft that require fluid only for manual operation of the wheel brakes to large transport aircraft where the systems are large and complex.

To achieve the necessary redundancy and reliability, the system may consist of several subsystems. Each subsystem has a power generating device (pump), reservoir, accumulator, heat exchanger, filtering system, etc. System operating pressure may vary from a couple hundred pounds per square inch (psi) in small aircraft and rotorcraft to 5,000 psi in large transports.

[Figure 4-5] Aircraft Hydraulic Systems

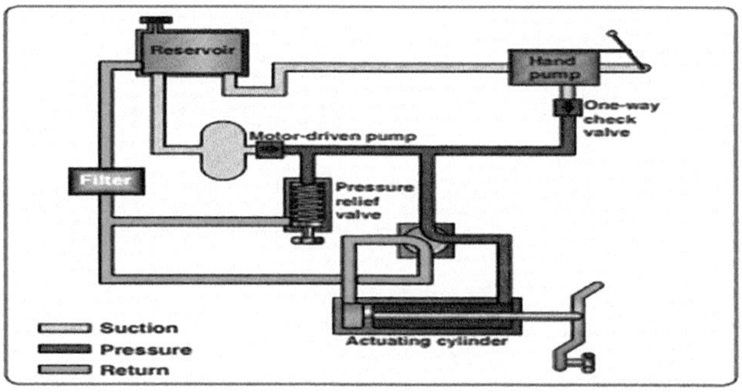

Words	accumulator 축전기 achieve 달성하다 complexity 복잡성 filtering system 필터링 시스템 fluid 유체 hydraulic 유압의 hydraulic brake system 유압 제동 시스템 heat exchanger 열 교환기 landing gear 착륙 장치 operate 작동시키다 pump 펌프 pressure 압력 redundancy 이중화 reservoir 저장소 reliability 신뢰성 subsystem 하위 시스템 transport 수송 wheel brake 바퀴 제동
Phrases	in order to ~하기 위해 provide a means for ~을 위한 수단을 제공하다

2) Hydraulic System Components

(1) Reservoirs

The reservoir is a tank in which an adequate supply of fluid for the system is stored. Fluid flows from the reservoir to the pump, where it is forced through the system and eventually returned to the reservoir. The reservoir not only supplies the operating needs of the system, but it also replenishes fluid lost through leakage. Furthermore, the reservoir serves as an overflow basin for excess fluid forced out of the system by thermal expansion (the increase of fluid volume caused by temperature changes), the accumulators, and by piston and rod displacement.

The reservoir also furnishes a place for the fluid to purge itself of air bubbles that may enter the system. Foreign matter picked up in the system may also be separated from the fluid in the reservoir or as it flows through line filters. Reservoirs are either pressurized or nonpressurized.

Words	accumulator 축압기 adequate 적절한 air bubbles 공기 방울 displacement 변위, 이동 eventually 결국 expansion 확장 fluid volume 액체 부피 foreign matter 이물질 furthermore 뿐만 아니라 leakage 누수 overflow 넘침 pressurized 압력이 있는 pump 펌프 purge 깨끗이하다, 제거하다 replenish 보충하다 reservoir 저장소 thermal expansion 열 팽창
Phrases	either A or B A 또는 B forced out of ～에서 밀려 나오다 furnish a place for ～을 위한 장소를 제공하다 not only A but also B A뿐만 아니라 B도 역시 separated from ～에서 분리되다

[Figure 4-6] Nonpressurized reservoir

[Figure 4-6] Nonpressurized reservoir   [Figure 4-7] Air-pressurized reservoir

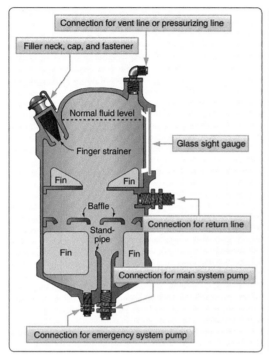

(2) Check Valve

A check valve allows fluid to flow unimpeded in one direction but prevents or restricts fluid flow in the opposite direction.

(3) Hydraulic Fuses

A hydraulic fuse is a safety device. Fuses may be installed at strategic locations throughout a hydraulic system. They detect a sudden increase in flow, such as a burst downstream, and shut off the fluid flow.

Words	**check valve** 체크 밸브 **detect** 감지하다 **hydraulic fuse** 유압 퓨즈 **location** 위치 **restrict** 제한하다 **shut off** 차단하다 **strategic** 전략적인 **unimpeded** 방해받지 않는
Phrases	**in one direction** 한 방향으로 **in the opposite direction** 반대 방향으로 **such as** ~와 같은

(4) Relief Valves

Hydraulic pressure must be regulated in order to use it to perform the desired tasks. A pressure relief valve is used to limit the amount of pressure being exerted on a confined liquid. This is necessary to prevent failure of components or rupture of hydraulic lines under excessive pressures. The pressure relief valve is, in effect, a system safety valve.

[Figure 4–8] Pressure relief valve

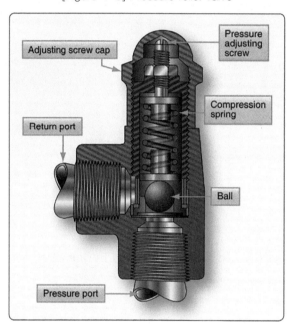

Words	components 부품　confined 제한된　exert 발휘하다　failure 고장 hydraulic 유압의　lines 라인　pressure 압력　prevent 방지하다 relief valve 압력 릴리프 밸브, 압력 해제 밸브　rupture 파열 safety valve 안전 밸브　system 시스템
Phrases	in effect 사실상　in order to ~ 위하여　limit the amount of ~의 양을 제한하다 under excessive pressures 과도한 압력 하에서

(5) Actuators

An actuating cylinder transforms energy in the form of fluid pressure into mechanical force, or action, to perform work. It is used to impart powered linear motion to some movable object or mechanism.

Actuating cylinders are of two major types: single action and double action. The single-action (single port) actuating cylinder is capable of producing powered movement in one direction only. The double-action (two ports) actuating cylinder is capable of producing powered movement in two directions.

[Figure 4-9] Linear actuator

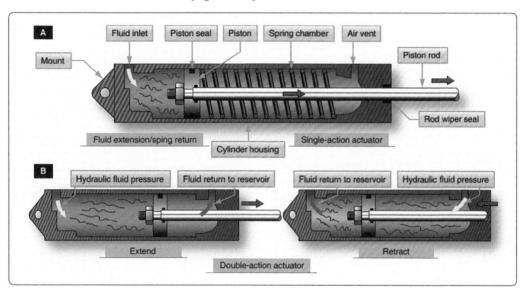

Words	**actuating** 구동하는 **cylinder** 실린더 **direction** 방향 **double action** 이중 작용 **fluid pressure** 유체 압력 **impart** 전달하다 **mechanical** 기계적인 **motion** 운동 **port** 포트 **powered** 동력화된 **single action** 단일 작용 **transform** 변형시키다
Phrases	**be capable of** ~할 수 있는 **in the form of** ~의 형태로 **in one direction** 한 방향으로 **in two directions** 두 방향으로

Multiple-Choice Questions

1. What is the main function of a hydraulic reservoir?

   a) To generate hydraulic pressure
   b) To store and supply fluid for the system
   c) To clean the fluid before use

2. What does a check valve do?

   a) Allows fluid to flow in both directions
   b) Stops fluid flow completely
   c) Allows fluid to flow in one direction only

3. Why are hydraulic fuses installed in a system?

   a) To prevent excessive fluid loss
   b) To increase fluid pressure
   c) To provide additional power

4. What is the primary purpose of a pressure relief valve?

   a) To reduce thermal expansion
   b) To prevent excessive pressure in the system
   c) To allow air bubbles to escape

5. Which type of hydraulic system is commonly found in small aircraft?

   a) Complex systems with subsystems
   b) Systems with high redundancy
   c) Systems requiring only wheel brake operation

6. What does a double-action actuator do?

   a) Transforms energy into mechanical force in one direction
   b) Transforms energy into mechanical force in two directions
   c) Releases excess fluid back to the reservoir

True/False (O/X) Questions

1. Hydraulic systems are only used for braking in modern aircraft. (O / X)

2. Hydraulic reservoirs store and supply fluid for the system. (O / X)

3. A check valve allows fluid to flow in any direction. (O / X)

4. Hydraulic fuses detect leaks and stop fluid flow. (O / X)

5. A hydraulic fuse helps prevent fluid loss during a burst. (O / X)

6. A relief valve ensures the system operates within safe pressure limits. (O / X)

Short Answer Questions

1. What is the primary role of a hydraulic system in an aircraft?

2. Name one component found in all hydraulic systems.

3. What type of actuator can produce motion in two directions?

4. Why is a pressure relief valve necessary in a hydraulic system?

5. What does a reservoir remove from the fluid?

6. Which hydraulic component converts fluid pressure into mechanical force?

Multiple-Choice Questions

1. b) To store and supply fluid for the system
2. c) Allows fluid to flow in one direction only
3. a) To prevent excessive fluid loss
4. b) To prevent excessive pressure in the system
5. c) Systems requiring only wheel brake operation
6. b) Transforms energy into mechanical force in two directions

True/False (O/X) Questions

1. X (Hydraulic systems operate landing gear, flight controls, and more, not just brakes.)
2. O
3. X (A check valve allows fluid to flow in one direction only.)
4. O
5. O
6. O

Short Answer Questions

1. To operate aircraft components such as landing gear, brakes, and flight controls.
2. Reservoir
3. Double-action actuator
4. To prevent excessive pressure that could damage components or lines.
5. Air bubbles and foreign matter
6. Actuator

4-3   Aircraft Pneumatic Systems

1) Introduction

Some aircraft manufacturers have equipped their aircraft with a high pressure pneumatic system (3,000 psi) in the past. The last aircraft to utilize this type of system was the Fokker F27. Such systems operate a great deal like hydraulic systems, except they employ air instead of a liquid for transmitting power.

Pneumatic systems are sometimes used for:

- Brakes
- Opening and closing doors
- Driving hydraulic pumps, alternators, starters, water injection pumps, etc.
- Operating emergency devices

Words	**alternators** 발전기 **brakes** 브레이크 **driving** 구동 **emergency** 비상 **except** 제외하다 **hydraulic** 유압 **pneumatics** 공기압 **starters** 스타터 **transmit** 전달하다
Phrases	**a great deal like** ～와 매우 유사하다 **equipped with** ～이 장착된 **instead of** ～의 대신으로 **in the past** 과거에 **sometimes used for** 때때로 ～에 사용된다

4-3 Aircraft Pneumatic Systems   61

[Figure 4-10] Pneumatic brake system

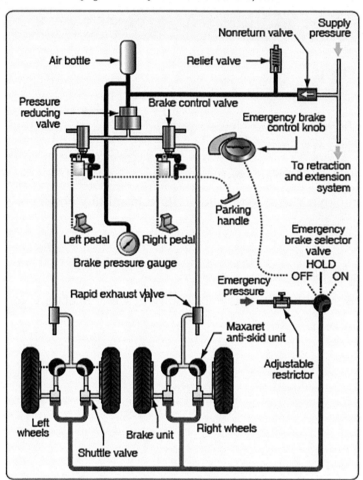

2) Pneumatic System Components

Pneumatic systems are often compared to hydraulic systems. Pneumatic systems do not utilize reservoirs, hand pumps, accumulators, regulators, or engine-driven or electrically driven power pumps for building normal pressure. But similarities do exist in some components.

(1) Air Compressors

On some aircraft, permanently installed air compressors have been added to recharge air bottles whenever pressure is used for operating a unit. Several types of compressors are used for this purpose. Some have two stages of compression, while others have three, depending on the maximum desired operating pressure.

(2) Relief Valves

Relief valves are used in pneumatic systems to prevent damage. They act as pressure limiting units and prevent excessive pressures from bursting lines and blowing out seals.

Words	**accumulator** 축압기　**air compressors** 공기 압축기　**blow** 타격, 불다　**burst** 파열, 부풀어 터지다　**electrical** 전기의　**engine-driven** 엔진 구동의　**pressure** 압력　**relief valves** 릴리프 밸브, 안전 밸브　**regulator** 조절기　**reservoirs** 저장소　**seals** 씰　**similarity** 유사성　**utilize** 활용하다　**valves** 밸브
Phrases	**compared to** ～와 비교하다　**for this purpose** 이 목적을 위해　**pressure limiting units** 압력 제한 장치　**used for** ～을 위해 사용되다

(3) Check Valves

Check valves are used in both hydraulic and pneumatic systems. Figure 4-11 illustrates a flap-type pneumatic check valve. Air enters the left port of the check valve, compresses a light spring, forcing the check valve open and allowing air to flow out the right port. But if air enters from the right, air pressure closes the valve, preventing a flow of air out the left port. Thus, a pneumatic check valve is a one-direction flow control valve.

[Figure 4-11] Flap-type pneumatic check valve

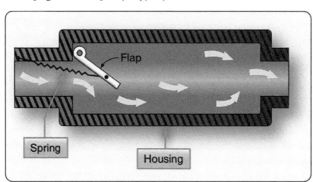

Words	air 공기 check valve 체크 밸브 compress 압축하다 hydraulic 유압의 pneumatic 공압의 port 포트 spring 스프링 valve 밸브
Phrases	be used in ~에서 사용되다 flow control 흐름 제어 in both 둘 다에서 one-direction 일방향 prevent from ~을 막다

(4) Restrictors

Restrictors are a type of control valve used in pneumatic systems. Figure 4-12 illustrates an orifice-type restrictor with a large inlet port and a small outlet port. The small outlet port reduces the rate of airflow and the speed of operation of an actuating unit.

(5) Filters

Pneumatic systems are protected against dirt by means of various types of filters. A micronic filter consists of a housing with two ports, a replaceable cartridge, and a relief valve.

[Figure 4-12] Pneumatic orifice valve

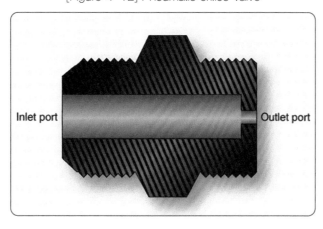

Inlet port   Outlet port

Words	**actuating unit** 작동 유닛 **cartridge** 카트리지 **dirt** 먼지 **filter** 필터 **inlet port** 입구 포트 **micronic filter** 마이크로닉 필터 **outlet port** 출구 포트 **pneumatic systems** 공압 시스템 **relief valve** 릴리프 밸브, 안전 밸브 **restrictor** 흐름제한장치
Phrases	**against dirt** 먼지로부터 **by means of** ∼를 사용하여 **consist of** ∼로 구성되다 **reduce the rate of** ∼의 속도를 줄이다

Aircraft Pneumatic Systems

Multiple-Choice Questions

1. What is the last aircraft model that used a high-pressure pneumatic system?

   a) Boeing 747
   b) Fokker F27
   c) Airbus A320

2. What is one common use of pneumatic systems in aircraft?

   a) Lighting systems
   b) Fuel injection
   c) Operating emergency devices

3. Which component is responsible for limiting pressure in pneumatic systems?

   a) Relief valve
   b) Check valve
   c) Restrictor

4. What is the main purpose of check valves in pneumatic systems?

   a) To regulate airflow speed
   b) To build pressure
   c) To allow one-directional flow of air

5. What type of restrictor is commonly used in pneumatic systems?

   a) Plate-type restrictor
   b) Orifice-type restrictor
   c) Valve-type restrictor

6. Where does air flow out in a pneumatic check valve?

   a) Left port only
   b) Right port only
   c) Both ports

True/False (O/X) Questions

1. Pneumatic systems use reservoirs and hand pumps to build pressure. (O / X)

2. Relief valves in pneumatic systems help prevent damage from excessive pressure. (O / X)

3. Check valves allow air to flow in both directions. (O / X)

4. Micronic filters prevent dirt from entering pneumatic systems. (O / X)

5. Restrictors in pneumatic systems reduce airflow speed. (O / X)

6. Hydraulic systems and pneumatic systems use the same types of compressors. (O / X)

Short Answer Questions

1. What substance do pneumatic systems use to transmit power?

2. Name one aircraft component operated by pneumatic systems.

3. What does a check valve prevent in pneumatic systems?

4. What is reduced by restrictors in pneumatic systems?

5. How many compression stages can some air compressors have?

6. What type of valve allows air to flow in one direction only?

Multiple-Choice Questions

1. b) Fokker F27

2. c) Operating emergency devices

3. a) Relief valve

4. c) To allow one-directional flow of air

5. b) Orifice-type restrictor

6. b) Right port only

True/False (O/X) Questions

1. X (Pneumatic systems do not use reservoirs or hand pumps for building normal pressure.)

2. O

3. X (Check valves allow air to flow in only one direction.)

4. O

5. O

6. X (Hydraulic and pneumatic systems use different types of compressors.)

Short Answer Questions

1. Air

2. Brakes (or other examples from the list: doors, emergency devices)

3. Reverse flow of air

4. Airflow speed

5. Two or three

6. Check valve

1) Introduction

Aircraft landing gear supports the entire weight of an aircraft during landing and ground operations. They are attached to primary structural members of the aircraft. The type of gear depends on the aircraft design and its intended use.

Most landing gear have wheels to facilitate operation to and from hard surfaces, such as airport runways. Other gear feature skids for this purpose, such as those found on helicopters, balloon gondolas, and in the tail area of some taildragger aircraft. Aircraft that operate to and from frozen lakes and snowy areas may be equipped with landing gear that have skis. Aircraft that operate to and from the surface of water have pontoon-type landing gear. Regardless of the type of landing gear utilized, shock absorbing equipment, brakes, retraction mechanisms, controls, warning devices, cowling, fairings, and structural members necessary to attach the gear to the aircraft are considered parts of the landing gear system.

[Figure 4–13] Basic landing gear types

Words	**absorbing** 흡수하는 **attached** 부착된 **brakes** 브레이크 **controls** 제어 장치 **equipment** 장비 **facilitate** 용이하게 하다 **fairings** 페어링 **features** 특징이 있다 **gear** 기어 **landing gear** 랜딩 기어 **pontoon** 플로트 **retraction** 수축 **skis** 스키 **structural members** 구조 구성 요소 **warning devices** 경고 장치 **wheels** 바퀴
Phrases	**attached to** ~에 부착된 **depends on** ~에 달려 있다 **necessary to** ~에 필요한 **regardless of** ~에 관계없이 **such as** ~와 같은 **to and from** ~에서 ~로

2) Landing Gear Arrangement

Three basic arrangements of landing gear are used: tail wheel-type landing gear (also known as conventional gear), tandem landing gear, and tricycle-type landing gear.

(1) Tail Wheel-Type Landing Gear

Tail wheel-type landing gear is also known as conventional gear because many early aircraft use this type of arrangement. The main gear are located forward of the center of gravity, causing the tail to require support from a third wheel assembly.

[Figure 4-14] The steerable tail wheel

Words	**arrangement** 배열, 배치 **center of gravity** 무게 중심 **conventional** 전통적인 **gear** 기어 **located** 위치한 **support** 지원 **tail** 꼬리 **third** 세 번째 **wheel** 바퀴
Phrases	**forward of** ~의 앞쪽에 **known as** ~로 알려진 **tail wheel-type landing gear** 꼬리바퀴식 착륙 장치

(2) Tandem Landing Gear

Few aircraft are designed with tandem landing gear. As the name implies, this type of landing gear has the main gear and tail gear aligned on the longitudinal axis of the aircraft. Sailplanes commonly use tandem gear, although many only have one actual gear forward on the fuselage with a skid under the tail.

[Figure 4-15] Tandem landing gear

Words	**aligned** 정렬된 **gear** 착륙 장치 **longitudinal** 종축의 **skid** 스키드 **tail** 꼬리 **tandem** 직렬의, 일렬의
Phrases	**forward on** ~의 앞에 **tandem landing gear** 직렬식 착륙 장치 **under the tail** 꼬리 아래에

(3) Tricycle–Type Landing Gear

The most commonly used landing gear arrangement is the tricycle-type landing gear. It is comprised of main gear and nose gear.

The tricycle-type landing gear arrangement consists of many parts and assemblies. These include air/oil shock struts, gear alignment units, support units, retraction and safety devices, steering systems, wheel and brake assemblies, etc.

[Figure 4–16] Tricycle–type landing gear

Words	air/oil shock struts 공기/오일 충격 스트럿 arrangement 배열, 배치 assembly 조립체 brake assemby 브레이크 조립체 commonly 일반적으로 gear alignment units 기어 정렬 장치 include ~을 포함하다 landing gear 착륙 장치 main gear 주 착륙 장치 nose gear 전방 착륙 장치 retraction devices 수축 장치, 접힘 장치 steering systems 조향 시스템 support units 지지 장치
Phrases	comprised of ~로 구성되다 consist of ~으로 구성되다

(4) Fixed and Retractable Landing Gear

Further classification of aircraft landing gear can be made into two categories: fixed and retractable. Many small, single-engine light aircraft have fixed landing gear, as do a few light twins. This means the gear is attached to the airframe and remains exposed to the slipstream as the aircraft is flown. As the speed of an aircraft increases, so does parasite drag. Mechanisms to retract and stow the landing gear to eliminate parasite drag add weight to the aircraft.

On slow aircraft, the penalty of this added weight is not overcome by the reduction of drag, so fixed gear is used. As the speed of the aircraft increases, the drag caused by the landing gear becomes greater and a means to retract the gear to eliminate parasite drag is required, despite the weight of the mechanism.

[Figure 4-17] Landing gear can be fixed or retractable

Words	eliminate 제거하다　fixed 고정된　landing gear 착륙 장치　mechanism 기계장치 overcome 극복하다　parasite drag 유해 항력 retractable 접어 넣을 수 있는, 수축 가능한　stow 보관하다 slipstream 슬립스트림 (비행기 뒤의 공기 흐름)
Phrases	as do ~와 마찬가지로　despite the weight of ~의 무게에도 불구하고 means to ~을 위한 방법　remain exposed to ~에 노출된 상태로 남다

3) Nose Wheel Steering Systems

The nose wheel on most aircraft is steerable from the flight deck via a nose wheel steering system. This allows the aircraft to be directed during ground operation.

(1) Small Aircraft

Most small aircraft have steering capabilities through the use of a simple system of mechanical linkages connected to the rudder pedals. Push-pull tubes are connected to pedal horns on the lower strut cylinder. As the pedals are depressed, the movement is transferred to the strut piston axle and wheel assembly which rotates to the left or right.

[Figure 4-18] Nose wheel steering on a light aircraft

Steering rod from rudder pedal

Words	depressed 눌린 linked 연결된 mechanical 기계적인 pedal 페달 strut 스트럿 steering 조향 transferred 전달된 wheel 휠
Phrases	connected to ~에 연결된 directed during ~동안 방향을 조정하다 in most 대부분에서 is steerable from ~에서 조향할 수 있다 move to ~로 이동하다

(2) Large Aircraft

Due to their mass and the need for positive control, large aircraft utilize a power source for nose wheel steering. Hydraulic power predominates. There are many different designs for large aircraft nose steering systems. Most share similar characteristics and components. Control of the steering is from the flight deck through the use of a small wheel, tiller, or joystick typically mounted on the left side wall.

[Figure 4–19] Hydraulic system flow diagram of large aircraft nose wheel steering system

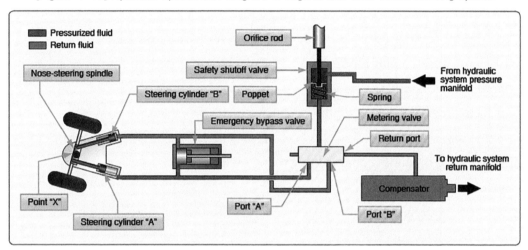

Words	**hydraulic** 유압의 **joystick** 조이스틱 **large** 대형의 **mass** 질량 **nose wheel** 전방 휠, 노즈 휠 **power source** 동력원 **predominates** 우세하다 **steering** 조향 **tiller** 조타기
Phrases	**be mounted on** ~에 장착되다 **due to** ~때문에 **in order to** ~하기 위해서 **through the use of** ~을 사용하여

Multiple-Choice Questions

1. What is the purpose of landing gear on an aircraft?

   a) To provide propulsion
   b) To support the aircraft during landing and ground operations
   c) To reduce drag during flight

2. Which type of landing gear is known as "conventional gear"?

   a) Tricycle-type landing gear
   b) Tandem landing gear
   c) Tail wheel-type landing gear

3. What type of landing gear is common on small, single-engine light aircraft?

   a) Retractable landing gear
   b) Fixed landing gear
   c) Skid landing gear

4. What controls nose wheel steering in most large aircraft?

   a) A tiller or joystick
   b) Rudder pedals
   c) A hydraulic valve

5. Which type of landing gear operates with skis?

   a) Tail wheel-type landing gear
   b) Fixed landing gear
   c) Gear for frozen lakes or snowy areas

6. How is steering typically controlled on small aircraft?

   a) Through hydraulic power
   b) Using rudder pedals connected to mechanical linkages
   c) By differential braking only

True/False (O/X) Questions

1. All aircraft landing gear have wheels.
   (O / X)

2. Tail wheel-type landing gear is also known as conventional gear. (O / X)

3. Tandem landing gear is commonly used in commercial airplanes. (O / X)

4. Tricycle-type landing gear is the most common arrangement for modern aircraft. (O / X)

5. Nose wheel steering in large aircraft is always powered by electricity. (O / X)

6. A tiller or joystick is commonly used for nose wheel steering in large aircraft. (O / X)

Short Answer Questions

1. What is the main purpose of landing gear?

2. Name one type of landing gear arrangement.

3. What is the most common landing gear arrangement for modern aircraft?

4. How is the landing gear classified based on movement?

5. How do small aircraft typically steer on the ground?

6. What system powers nose wheel steering in large aircraft?

Multiple-Choice Questions

1. b) To support the aircraft during landing and ground operations
2. c) Tail wheel-type landing gear
3. b) Fixed landing gear
4. a) A tiller or joystick
5. c) Gear for frozen lakes or snowy areas
6. b) Using rudder pedals connected to mechanical linkages

True/False (O/X) Questions

1. X (Some landing gear use skis or pontoons instead of wheels.)
2. O
3. X (Tandem landing gear is rarely used, commonly seen on sailplanes.)
4. O
5. X (Hydraulic systems are commonly used for nose wheel steering in large aircraft.)
6. O

Short Answer Questions

1. To support the aircraft during landing and ground operations.
2. Tail wheel-type, tricycle-type, or tandem.
3. Tricycle-type landing gear.
4. Fixed or retractable.
5. By rudder pedals and mechanical linkages.
6. Hydraulic power.

4-5   Aircraft Fuel Systems

◁» MP3 4-5-1

1) Types of Aviation Fuel

Each aircraft engine is designed to burn a certain fuel. Use only the fuel specified by the manufacturer. Mixing fuels is not permitted. There are two basic types of fuel discussed in this section: reciprocating-engine fuel (also known as gasoline or AV-GAS) and turbine-engine fuel (also known as jet fuel or kerosene).

(1) Reciprocating Engine Fuel

Reciprocating engines burn gasoline, also known as AVGAS(Aviation gasoline). It is specially formulated for use in aircraft engines. Combustion releases energy in the fuel, which is converted into the mechanical motion of the engine. AVGAS of any variety is primarily a hydrocarbon compound refined from crude oil by fractional distillation. Aviation gasoline is different from the fuel refined for use in turbine-powered aircraft. AVGAS is very volatile and extremely flammable, with a low flash point. Turbine fuel is a kerosene-type fuel with a much higher flash point, so it is less flammable.

Words	aviation gasoline 항공용 가솔린 (AVGAS) burn 연소하다 combustion 연소 convert 전환하다 crude oil 원유 flammable 가연성 formulate 공식화하다 fuel 연료 flash point 인화점 hydrocarbon 탄화수소 jet fuel 제트 연료 kerosene 등유 mixing 혼합 reciprocating engine 왕복엔진 refined 세련된 turbine engine 터빈 엔진 volatile 휘발성
Phrases	flash point 인화점 fractional distillation 분별 증류 specified by ~에 의해 지정된

(2) Turbine Engine Fuel

Aircraft with turbine engines use a type of fuel different from that of reciprocating aircraft engines. Commonly known as jet fuel, turbine engine fuel is designed for use in turbine engines and should never be mixed with aviation gasoline or introduced into the fuel system of a reciprocating aircraft engine fuel system.

[Figure 4-20] Basic Fuel System

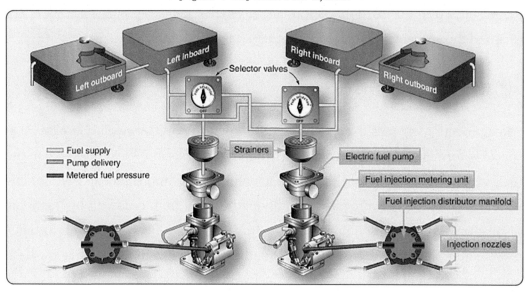

Words	**aviation** 항공의 **commonly** 흔히, 보통 **gasoline** 가솔린 **jet fuel** 제트 연료 **reciprocating** 왕복식의 **system** 시스템 **turbine** 터빈
Phrases	**commonly known as** ~로 일반적으로 알려져 있는 **designed for use in** ~에서 사용되도록 설계된 **introduced into** ~에 도입되다

2) Aircraft Fuel Systems

Each aircraft fuel system must store and deliver clean fuel to the engines at a pressure and flow rate able to sustain operations regardless of the operating conditions of the aircraft.

(1) Small Single-Engine Aircraft

Small single-engine aircraft fuel systems vary depending on factors, such as tank location and method of metering fuel to the engine. A high-wing aircraft fuel system can be designed differently from one on a low-wing aircraft. An aircraft engine with a carburetor has a different fuel system than one with fuel injection.

[Figure 4-21] Single reciprocating engine aircraft with fuel tanks

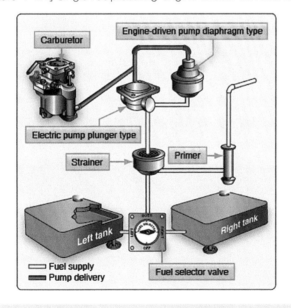

Words	**carburetor** 기화기(카뷰레터) **fuel injection** 연료 분사 **fuel system** 연료 시스템 **high-wing** 고익 날개 **low-wing** 저익 날개 **location** 위치 **metering** 계량 **pressure** 압력 **single-engine** 단발 **sustain** 지속시키다 **system** 시스템 **vary** 서로 다르다
Phrases	**able to** ~할 수 있는 **depending on** ~에 따라 **regardless of** ~에 관계없이 **similar to** ~와 유사한

(2) Large Reciprocating-Engine Aircraft

Large, multi-engine transport aircraft powered by reciprocating radial engines are no longer produced. However, many are still in operation. They are mostly carbureted and share many features with the light aircraft systems previously discussed.

Words	carbureted 기화기가 장착된 in operation 운항 중인 large 대형의 multiengine 다발 엔진 operation 운항 radial 성형, 방사형 reciprocating 왕복식 share 함께 쓰다 transport 수송
Phrases	no longer 더 이상 ~하지 않다 powered by ~로 구동되다

(3) Jet Transport Aircraft

Fuel systems on large transport category jet aircraft are complex with some features and components not found in reciprocating-engine aircraft fuel systems. They typically contain more redundancy and facilitate numerous options from which the crew can choose while managing the aircraft's fuel load. Features like an onboard APU, single point pressure refueling, and fuel jettison systems, which are not needed on smaller aircraft, add to the complexity of an airliner fuel system.

Most transport category aircraft fuel systems are very much alike. Integral fuel tanks are the norm with much of each wing's structure sealed to enable its use as a fuel tank. Center wing section or fuselage tanks are also common. These may be sealed structure or bladder type. Jet transport aircraft carry tens of thousands of pounds of fuel on board. Figure 4-22 shows a diagram of a Boeing 777 fuel tank configuration with tank capacities.

[Figure 4-22] Boeing 777 fuel tank configuration

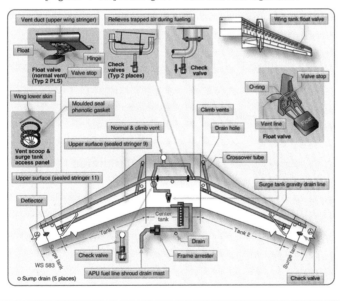

Words	**airliner** 여객기 **alike** 비슷한 **APU** 보조동력장치 **bladder** 고무형(브래더) **complexity** 복잡성 **configuration** 배치 **facilitate** 가능하게 하다 **fuel load** 연료 부하 **fuselage** 동체 **in the norm** 표준으로 **on board** 기내에 **redundancy** 중복 **transport** 수송 **typically** 보통, 일반적으로
Phrases	**add to** ~을 더하다 **carry on board** 탑재하다 **not found in** ~에서 찾을 수 없는

Aircraft Fuel Systems

Multiple-Choice Questions

1. What type of fuel is used in reciprocating engines?

   a) AVGAS
   b) Jet fuel
   c) Diesel fuel

2. What is the primary difference between AVGAS and jet fuel?

   a) Jet fuel is refined from natural gas.
   b) AVGAS has a lower flash point than
   jet fuel.
   c) AVGAS is used only in turbine
   engines.

3. Which type of tank is commonly used
in jet transport aircraft?

   a) Auxiliary tanks
   b) External tanks
   c) Integral fuel tanks.

4. Where is the fuel typically stored in
transport category aircraft?

   a) Vertical stabilizer
   b) Wings and center wing section
   c) Landing gear compartments

5. Why should AVGAS and jet fuel not
be mixed?

   a) Mixing causes engine damage.
   b) They have different energy content.
   c) They are chemically incompatible.

6. What type of fuel system do most
multiengine transport aircraft use?

   a) Gravity-fed systems
   b) Carbureted systems
   c) Complex jet fuel systems

True/False (O/X) Questions

1. AVGAS is specifically designed for turbine engines. (O / X)

2. Jet fuel has a higher flash point than AVGAS. (O / X)

3. Fuel jettison systems are commonly used in small, single-engine aircraft. (O / X)

4. Integral fuel tanks are typically found in jet transport aircraft. (O / X)

5. AVGAS is less volatile than jet fuel. (O / X)

6. Mixing AVGAS with jet fuel is allowed under certain conditions. (O / X)

Short Answer Questions

1. What does AVGAS stand for?

2. Which type of engine uses jet fuel?

3. Where is the refueling station typically located on large transport aircraft?

4. Why is AVGAS considered highly flammable?

5. What is the purpose of an APU in jet transport aircraft?

6. Where are integral fuel tanks commonly located?

Multiple-Choice Questions

1. a) AVGAS
2. b) AVGAS has a lower flash point than jet fuel.
3. c) Integral fuel tanks
4. b) Wings and center wing section
5. a) Mixing causes engine damage.
6. c) Complex jet fuel systems

True/False (O/X) Questions

1. X (AVGAS is designed for reciprocating engines, not turbine engines.)
2. O
3. X (Fuel jettison systems are used in large transport aircraft, not small aircraft.)
4. O
5. X (AVGAS is more volatile than jet fuel.)
6. X (Mixing fuels is strictly prohibited.)

Short Answer Questions

1. Aviation Gasoline
2. Turbine engines
3. On the wings
4. It has a low flash point.
5. To provide power for systems without using the main engines
6. Wings or fuselage

4-6 Aircraft Lighting Systems

◁) MP3 4-6-1

1) Introduction

Aircraft lighting systems provide illumination for both exterior and interior use. Lights on the exterior provide illumination for such operations as landing at night, inspection of icing conditions, and safety from midair collision. Interior lighting provides illumination for instruments, cockpits, cabins, and other sections occupied by crewmembers and passengers. Certain special lights, such as indicator and warning lights, indicate the operation status of equipment.

Words	cockpit 조종석 collision 충돌 crew member 승무원 equipment 장비 exterior 외부 illumination 조명 indicator 지시등 interior 내부 midair collision 공중 충돌 operation 작업, 운영 passenger 승객 status 상태 warning light 경고등
Phrases	both exterior and interior 내외부 모두 occupied by ~에 의해 점유된 provide illumination for ~에 조명을 제공하다 such as ~와 같은 safety from ~로부터의 안전

2) Exterior Lights

Position, anticollision, landing, and taxi lights are common examples of aircraft exterior lights. Some lights are required for night operations. Other types of exterior lights, such as wing inspection lights, are of great benefit for specialized flying operations.

[Figure 4–23] A left wing tip position light (red) and a white strobe light

Words	**aircraft** 항공기 **anticollision** 충돌 방지 **benefit** 혜택 **exterior** 외부 **inspection** 점검 **landing light** 착륙등 **position light** 위치(표시)등 **taxi light** 유도등
Phrases	**such as** ∼와 같은 **required for** ∼에 필수적인

(1) Position Lights

Aircraft operating at night must be equipped with position lights that meet the minimum requirements specified by Title 14 of the Code of Federal Regulations. A set of position lights consist of one red, one green, and one white light.

The green light unit is always mounted at the extreme tip of the right wing. The red unit is mounted in a similar position on the left wing. The white unit is usually located on the vertical stabilizer in a position where it is clearly visible through a wide angle from the rear of the aircraft.

[Figure 4-24] A right wing tip position light

Words	**angle** 각도 **at night** 야간에 **equipped** 장착된 **extreme tip** 끝단 **minimum requirement** 최소 요구 사항 **position light** 위치표시등 **regulation** 규정 **stabilizer** 안정판 **visible** 보이는
Phrases	**be equipped with** ～을 갖추고 있다 **be mounted on** ～에 장착되다 **consist of** ～로 구성되다 **in a position where** ～한 위치에

(2) Anticollision Lights

An anticollision light system may consist of one or more lights. They are rotating beam lights that are usually installed on top of the fuselage or tail in such a location that the light does not affect the vision of the crew member or detract from the visibility of the position lights. Large transport type aircraft use an anticollision light on top and one on the bottom of the aircraft.

[Figure 4-25] Anticollision lights

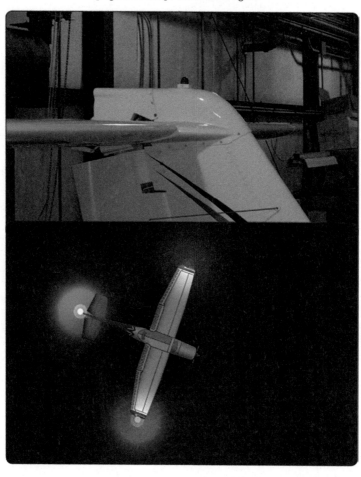

Words	**affect** 영향을 미치다　**anticollision** 충돌 방지　**beam** 빔　**crew member** 승무원 **illuminate** 비추다　**installed** 설치된　**light system** 조명 시스템　**rotating** 회전하는 **visibility** 가시성
Phrases	**detract from** ～을 저해하다　**in such a location** 그러한 위치에 **on top of** ～의 상부에

(3) Landing Lights

Landing lights are installed in aircraft to illuminate runways during night landings. These lights are very powerful and are directed by a parabolic reflector at an angle providing a maximum range of illumination.

Landing lights of smaller aircraft are usually located midway in the leading edge of each wing or streamlined into the aircraft surface. Landing lights for larger transport category aircraft are usually located in the leading edge of the wing close to the fuselage.

[Figure 4-26] Landing lights

Words	**leading edge** 앞전　**landing lights** 착륙등　**midway** 중간, 도중에 **parabolic reflector** 포물선 반사체　**powerful** 강력한　**runway** 활주로 **streamlined** 매끄럽게 통합된
Phrases	**close to** ~에 가까운　**midway in** ~의 중간에 **streamline into** ~에 매끄럽게 통합되다

(4) Taxi Lights

Taxi lights are designed to provide illumination on the ground while taxiing or towing the aircraft to or from a runway, taxi strip, or in the hangar area. [Figure 4-27] Taxi lights are not designed to provide the degree of illumination necessary for landing lights. On aircraft with tricycle landing gear, either single or multiple taxi lights are often mounted on the non-steerable part of the nose landing gear.

[Figure 4-27] Taxi lights

Words	**degree** 정도 **hangar** 격납고 **illumination** 조명 **multiple** 복수의 **non-steerable** 조향되지 않는 **runway** 활주로 **taxi light** 유도등 **taxi strip** 유도로 **towing** 견인
Phrases	**be designed to** ~하도록 설계되다 **be mounted on** ~에 장착되다 **either single or multiple** 단일 또는 복수의 **necessary for** ~에 필요한

3) Interior Lights

Aircraft are equipped with interior lights to illuminate the cabin. [Figure 4-28] Often white and red light settings are provided. Commercial aircraft have a lighting systems that illuminates the main cabin, an independent lighting system so that passengers can read when the cabin lights are off, and an emergency lighting system on the floor of the aircraft to aid passengers of the aircraft during an emergency.

[Figure 4-28] Interior cockpit and cabin light system

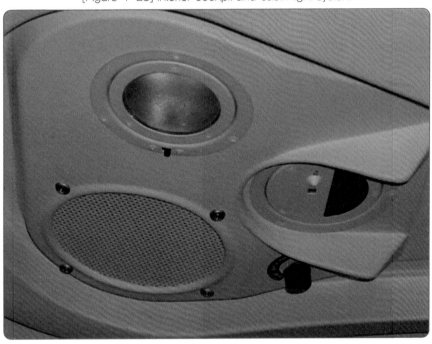

Words	**commercial** 상업용 **emergency** 비상 **illuminate** 밝히다 **independent** 독립적인 **interior lights** 내부 조명 **passengers** 승객 **settings** 설정
Phrases	**be equipped with** ~을 갖추고 있다. **on the floor of** ~의 바닥에

Multiple-Choice Questions

1. What is the main purpose of aircraft lighting systems?

   a) To decorate the aircraft
   b) To provide illumination for exterior and interior operations
   c) To increase the aircraft's speed

2. Which light is mounted at the extreme tip of the right wing?

   a) Green light
   b) Red light
   c) White light

3. What is the main function of anticollision lights?

   a) To illuminate runways during night landings
   b) To indicate the operation status of equipment
   c) To increase visibility and prevent midair collisions

4. What does the emergency lighting system in the cabin help passengers do?

   a) See the cockpit
   b) Exit the aircraft during an emergency
   c) Read when cabin lights are off

5. What is the main purpose of position lights?

   a) To help the crew see in the cockpit
   b) To indicate the aircraft's position during night operations
   c) To illuminate the runway during landing

6. Which of the following is NOT a type of interior lighting?

   a) Main cabin lights
   b) Reading lights for passengers
   c) Taxi lights

True/False (O/X) Questions

1. Position lights consist of one red, one green, and one white light. (O / X)

2. Landing lights are used for ground operations like taxiing. (O / X)

3. Position lights are required for aircraft operating during the day. (O / X)

4. Landing lights are usually located on the aircraft's wings close to the fuselage. (O / X)

5. The anticollision light system consists of only one light. (O / X)

6. Emergency lighting systems are usually mounted in the aircraft cabin to help passengers see during an emergency. (O / X)

Short Answer Questions

1. What is the purpose of exterior lights on an aircraft?

2. Where are the green position lights mounted on an aircraft?

3. What type of light is used to prevent midair collisions?

4. What type of lighting is used to illuminate the cabin during an emergency?

5. What color is the position light on the left wing?

6. Where are landing lights typically located on larger transport aircraft?

Multiple-Choice Questions

1. b) To provide illumination for exterior and interior operations
2. a) Green light
3. c) To increase visibility and prevent midair collisions
4. b) Exit the aircraft during an emergency
5. b) To indicate the aircraft's position during night operations
6. c) Taxi lights

True/False (O/X) Questions

1. O
2. X (Landing lights are used for night landings, not ground operations.)
3. X (Position lights are required for night operations, not daytime.)
4. O
5. X (Large aircraft typically have multiple anticollision lights.)
6. O

Short Answer Questions

1. To provide illumination for both exterior and interior operations, including night landings and crew member visibility.
2. On the extreme tip of the right wing.
3. Anticollision lights
4. The emergency lighting system
5. Red
6. In the leading edge of the wing, close to the fuselage.

CHAPTER 5
Aircraft Engines

5-1 General

Aircraft engines can be classified by several methods. They can be classed by operating cycles, cylinder arrangement, or the method of thrust production. All are heat engines that convert fuel into heat energy that is converted to mechanical energy to produce thrust.

Most of the current aircraft engines are of the internal combustion type because the combustion process takes place inside the engine. Aircraft engines come in many different types, such as gas turbine based, reciprocating piston, rotary, two or four cycle, spark ignition, diesel, and air or water cooled. Reciprocating and gas turbine engines also have subdivisions based on the type of cylinder arrangement (piston) and speed range (gas turbine).

[Figure 5-1] Types of engines

Words	**arrangement** 배열 **combustion** 연소 **convert** 변환하다 **cycle** 사이클 **diesel** 디젤 **energy** 에너지 **engine** 엔진 **internal combustion** 내연 **mechanical** 기계적인 **reciprocating** 왕복(왕복식) **rotary** 로터리 **spark ignition** 스파크 점화, 불꽃 점화 **thrust** 추력 **turbine** 터빈
Phrases	**based on** ~에 기반하여 **be classified by** ~로 분류되다 **convert to** ~로 변환하다 **divided into** ~로 나뉘다 **take place** 발생하다

5-2 Reciprocating Engines

🔊 MP3 5-2

The basic major components of a reciprocating engine are the crankcase, cylinders, pistons, connecting rods, valves, valve-operating mechanism, and crankshaft. In the head of each cylinder are the valves and spark plugs. One of the valves is in a passage leading from the induction system; the other is in a passage leading to the exhaust system. Inside each cylinder is a movable piston connected to a crankshaft by a connecting rod.

[Figure 5-2] A typical four-cylinder opposed engine

Words	components 요소, 부품 connecting rod 커넥팅 로드 crankcase 크랭크케이스 crankshaft 크랭크축 cylinder 실린더 exhaust system 배기 시스템 induction system 흡입 시스템 movable 움직이는 passage 통로, 관 piston 피스톤 spark plug 점화 플러그, 스파크 플러그 valve 밸브 valve-operating mechanism 밸브 작동 메커니즘
Phrases	connected to ~에 연결된 in a passage leading to ~로 이어지는 통로에 in the head of ~의 상단에

Gas Turbine Engines

In a reciprocating engine, the functions of intake, compression, combustion, and exhaust all take place in the same combustion chamber. Consequently, each must have exclusive occupancy of the chamber during its respective part of the combustion cycle. A significant feature of the gas turbine engine is that separate sections are devoted to each function, and all functions are performed simultaneously without interruption.

A typical gas turbine engine consists of air inlet, compressor section, combustion section, turbine section, exhaust section, accessory section. The systems necessary for starting, lubrication, fuel supply, and auxiliary purposes, such as anti-icing, cooling, and pressurization.

Words	air inlet 공기 흡입구 anti-icing 제빙 combustion chamber 연소실 combustion cycle 연소 사이클 combustion section 연소 구역 compressor section 압축기 구역 consequently 따라서 exclusive 독점적인 exhaust section 배기 구역 gas turbine engine 가스터빈 엔진 lubrication 윤활 occupancy 점유 perform 수행하다 pressurization 압력 유지 respective 각각의 simultaneously 동시에 turbine section 터빈 구역
Phrases	necessary for ~에 필요하다

[Figure 5–3] Gas turbine engines

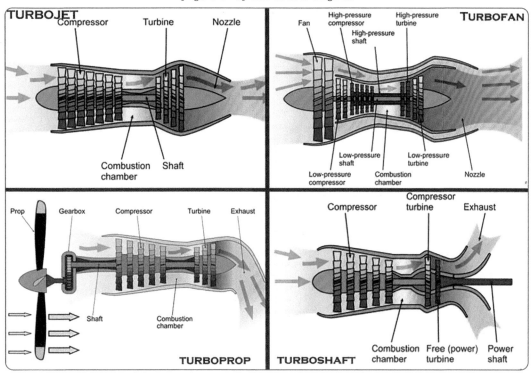

Multiple-Choice Questions

1. What is one of the primary methods for classifying aircraft engines?

   a) Operating cycle
   b) Fuel type
   c) Number of blades

2. What does a reciprocating engine primarily convert into mechanical energy?

   a) Air
   b) Fuel
   c) Water

3. Where are the valves and spark plugs located in a reciprocating engine?

   a) In the crankcase
   b) In the cylinder head
   c) In the exhaust system

4. Which component connects the piston to the crankshaft in a reciprocating engine?

   a) Valve
   b) Crankcase
   c) Connecting rod

5. Which of the following sections is part of a typical gas turbine engine?

   a) Combustion section
   b) Fuel reservoir
   c) Air filtration system

6. What is the function of the turbine section in a gas turbine engine?

   a) To cool the engine
   b) To provide power to drive the compressor
   c) To add fuel to the system

True/False (O/X) Questions

1. The crankcase is a major component of a gas turbine engine. (O/X)

2. The gas turbine engine has separate sections for each function, unlike the reciprocating engine. (O/X)

3. The piston in a reciprocating engine is connected to the crankshaft by a connecting rod. (O/X)

4. The exhaust section is responsible for cooling the engine in a gas turbine engine. (O/X)

5. Reciprocating engines usually use a four-stroke cycle. (O/X)

6. In a gas turbine engine, the compressor section supplies air to the combustion section. (O/X)

Short Answer Questions

1. What does a gas turbine engine do with air and fuel?

2. Where are the valves located in a reciprocating engine?

3. What is the main function of the compressor section in a gas turbine engine?

4. Which type of engine is primarily used in most modern aircraft?

5. What is the primary purpose of the turbine section in a gas turbine engine?

6. Where is the spark plug located in a reciprocating engine?

Multiple-Choice Questions

1. a) Operating cycle
2. b) Fuel
3. b) In the cylinder head
4. c) Connecting rod
5. a) Combustion section
6. b) To provide power to drive the compressor

True/False (O/X) Questions

1. X (The crankcase is a major component of a reciprocating engine, not a gas turbine engine.)
2. O
3. O
4. X (The exhaust section primarily expels gases and does not cool the engine; that is done by the cooling system.)
5. O
6. O

Short Answer Questions

1. The air and fuel are mixed in the combustion section of a gas turbine engine for combustion.
2. The valves are located in the cylinder head.
3. The compressor section compresses air to supply to the combustion section.
4. The gas turbine engine is primarily used in modern aircraft.
5. The turbine section drives the compressor and produces power.
6. The spark plug is located in the cylinder head.

CHAPTER 6
Propellers

The propeller, the unit that must absorb the power output of the engine, has passed through many stages of development. Although most propellers are two-bladed, great increases in power output have resulted in the development of four- and six-bladed propellers of large diameters. There are several forces acting on the propeller as it turns; a major one is centrifugal force. This force at high rpm tends to pull the blades out of the hub, so blade weight is very important to the design of a propeller.

As aircraft speeds increased, turbofan engines were used for higher speed aircraft. Propeller-driven aircraft have several advantages and are widely used for applications in turboprops and reciprocating engine installations. Takeoff and landing can be shorter and less expensive. New blade materials and manufacturing techniques have increased the efficiency of propellers. Many smaller aircraft will continue to use propellers well into the future.

Words	**absorb** 흡수하다 **centrifugal force** 원심력 **diameter** 직경 **efficiency** 효율성 **force** 힘 **hub** 허브 **material** 재료 **power output** 출력 동력 **propeller** 프로펠러 **speed** 속도
Phrases	**acting on** ～에 작용하다 **in the future** 미래에 **resulted in** 그 결과 ～가 되다 **tend to** ～ 하는 경향이 있다

[Figure 6-1] General of propeller

1) Pusher Propellers

Pusher propellers are those mounted on the downstream end of a drive shaft behind the supporting structure. Pusher propellers are constructed as fixed- or variable-pitch propellers. Seaplanes and amphibious aircraft have used a greater percentage of pusher propellers than other kinds of aircraft.

[Figure 6-2] Pusher propellers

Words	**amphibious** 수륙 양용의 **constructed** 제작된 **drive shaft** 구동 샤프트 **end** 끝 **fixed-pitch** 고정 피치 **pusher propellers** 추진식 프로펠러(푸셔 프로펠러) **propellers** 프로펠러 **supporting structure** 지지 구조물 **variable-pitch** 가변 피치
Phrases	**be constructed as** ~로 구성되어 있다 **be mounted on** ~에 장착되다 **downstream end** 하류 끝

2) Fixed-Pitch Propeller

Generally, this type of propeller is one piece and is constructed of wood or aluminum alloy.

Fixed-pitch propellers are designed for best efficiency at one rotational and forward speed. They are designed to fit a set of conditions of both airplane and engine speeds and any change in these conditions reduces the efficiency of both the propeller and the engine.

The fixed-pitch propeller is used on airplanes of low power, speed, range, or altitude. Many single-engine aircraft use fixed-pitch propellers and the advantages to these are less expense and their simple operation. This type of propeller does not require any control inputs from the pilot in flight.

[Figure 6–3] Fixed–pitch propeller

Words	**advantages** 장점 **alloy** 합금 **altitude** 고도 **expense** 비용 **fixed-pitch** 고정 피치 **flight** 비행 **forward** 전방의 **range** 비행 범위 **rotational** 회전의
Phrases	**be designed for** ～로 설계되다 **be used on** ～에 사용되다 **in flight** 비행 중

Multiple-Choice Questions

1. What is a major force acting on a propeller during operation?

   a) Gravity
   b) Air resistance
   c) Centrifugal force

2. What is the typical number of blades on most propellers?

   a) One
   b) Two
   c) Four

3. What is a key characteristic of fixed-pitch propellers?

   a) They can change their pitch during flight
   b) They are designed for a specific rotational and forward speed
   c) They are used only on high-speed aircraft

4. Which type of aircraft commonly uses pusher propellers?

   a) Seaplanes and amphibious aircraft
   b) Commercial airliners
   c) Helicopters

5. Which type of propeller does not require control inputs from the pilot in flight?

   a) Variable-pitch propeller
   b) Fixed-pitch propeller
   c) Feathering propeller

6. What is the common material used in fixed-pitch propellers?

   a) Plastic
   b) Titanium
   c) Wood or aluminum alloy

True/False (O/X) Questions

1. The centrifugal force on the propeller blades helps to keep them in place at high rpm. (O / X)

2. Seaplanes and amphibious aircraft rarely use pusher propellers. (O / X)

3. Fixed-pitch propellers require control inputs from the pilot during flight. (O / X)

4. Turbofan engines are typically used in propeller-driven aircraft. (O / X)

5. Four and six-bladed propellers are often used on high-power engines. (O / X)

6. Propeller-driven aircraft have the advantage of longer takeoff and landing distances. (O / X)

Short Answer Questions

1. What is the primary function of a propeller?

2. What type of propeller is mounted behind the supporting structure?

3. What type of aircraft uses more pusher propellers?

4. How many blades do most propellers typically have?

5. Which type of propeller is designed for best efficiency at a set of conditions?

6. What has increased the efficiency of propellers in recent years?

Propellers

Multiple-Choice Questions

1. c) Centrifugal force

2. b) Two

3. b) They are designed for a specific rotational and forward speed

4. a) Seaplanes and amphibious aircraft

5. b) Fixed-pitch propeller

6. c) Wood or aluminum alloy

True/False (O/X) Questions

1. O

2. X (Seaplanes and amphibious aircraft use a greater percentage of pusher propellers.)

3. X (Fixed-pitch propellers do not require control inputs from the pilot.)

4. X (Turbofan engines are used for high-speed aircraft, not typically in propeller-driven aircraft.)

5. O

6. X (Propeller-driven aircraft have shorter and less expensive takeoff and landing distances.)

Short Answer Questions

1. To absorb the power output of the engine.

2. Pusher propeller

3. Seaplanes and amphibious aircraft.

4. Two

5. Fixed-pitch propeller

6. New blade materials and manufacturing techniques.

Appendix

Text Translation

Chapter 1
Aircraft Structures

1-1. General

The airframe of a fixed-wing aircraft consists of five principal units: the fuselage, wings, stabilizers, flight control surfaces, and landing gear.

고정익 항공기의 기체는 5개의 주요 구성 부품으로 이루어진다: 동체, 날개, 안정판, 비행 조종 면, 그리고 착륙 장치.

Airframe structural components are constructed from a wide variety of materials.

기체 구조 부품은 매우 다양한 재료로 제작된다.

The earliest aircraft were constructed primarily of wood.

초기 항공기는 주로 나무로 제작되었다.

Steel tubing and the most common material, aluminum, followed.

강철 튜브와 가장 일반적인 재료인 알루미늄이 뒤이어 사용되었다.

Many newly certified aircraft are built from molded composite materials, such as carbon fiber.

새로 인증된 많은 항공기는 탄소 섬유와 같은 성형 복합 재료로 제작된다.

Structural members of an aircraft's fuselage include stringers, longerons, ribs, bulkheads, and more.

항공기의 동체 구조 부재에는 스트링거, 롱거론, 리브, 벌크헤드 등이 포함된다.

The main structural member in a wing is called the wing spar.

날개의 주요 구조 부재는 날개 스파라고 불린다.

The skin of aircraft can also be made from a variety of materials, ranging from impregnated fabric to plywood, aluminum, or composites.

항공기의 외피는 함침 직물에서 합판, 알루미늄, 또는 복합 재료에 이르기까지 다양한 재료로 만들어질 수 있다.

The entire airframe and its components are joined by rivets, bolts, screws, and other fasteners.

전체 기체와 그 구성 요소는 리벳, 볼트, 나사 및 기타 체결 장치로 결합된다.

Welding, adhesives, and special bonding techniques are also used.

용접, 접착제, 그리고 특수 접합 기술도 사용된다.

1-2. Fixed-Wing Aircraft

1) Fuselage

The fuselage is the main structure or body of the fixed-wing aircraft.

동체는 고정익 항공기의 주요 구조물 또는 본체이다.

It provides space for cargo, controls, accessories, passengers, and other equipment.

동체는 화물, 조종 장치, 부속품, 승객, 및 기타 장비를 위한 공간을 제공한다.

In single-engine aircraft, the fuselage houses the powerplant.

단발 엔진 항공기에서는 동체가 동력 장치를 수용한다.

In multiengine aircraft, the engines may be either in the fuselage, attached to the fuselage, or suspended from the wing structure.

다발 엔진 항공기에서는 엔진이 동체 내부에 있거나, 동체에 부착되거나, 날개 구조물에 매달릴 수 있다.

There are two general types of fuselage construction: truss and monocoque.

동체 구조에는 두 가지 일반적인 유형이 있다: 트러스형과 모노코크형이다.

2) Wing Configurations

Wings are airfoils that, when moved rapidly through the air, create lift.

날개는 공기 중에서 빠르게 움직일 때 양력을 생성하는 에어포일이다.

They are built in many shapes and sizes.

날개는 다양한 모양과 크기로 제작된다.

Wing design can vary to provide certain desirable flight characteristics.

날개 설계는 특정한 비행 특성을 제공하기 위해 다양하게 변경될 수 있다.

Control at various operating speeds, the amount of lift generated, balance, and stability all change as the shape of the wing is altered.

다양한 작동 속도에서의 제어, 생성된 양력의 양, 균형 및 안정성은 날개 모양이 변경될 때 모두 달라진다.

The wing tip may be square, rounded, or even pointed.

날개의 끝은 네모나거나, 둥글거나, 또는 뾰족할 수 있다.

Figure 1-4 shows a number of typical wing leading and trailing edge shapes.

그림 1-4는 여러 가지 일반적인 날개 앞전과 뒷전 모양을 보여준다.

3) Empennage

The empennage of an aircraft is also known as the tail section.

항공기의 엠퍼너지는 꼬리 부분으로도 알려져 있다.

Most empennage designs consist of a tail cone, fixed aerodynamic surfaces or stabilizers, and movable aerodynamic surfaces.

대부분의 엠퍼너지 설계는 꼬리 콘, 고정된 공기역학적 표면 또는 안정판, 그리고 가동식 공기역학적 표면으로 구성된다.

The tail cone serves to close and streamline the aft end of most fuselages.

꼬리 콘은 대부분 동체의 후미를 닫고 유선형으로 만드는 역할을 한다.

The cone is made up of structural members like those of the fuselage; however, cones are usually of lighter construction since they receive less stress than the fuselage.

이 콘은 동체와 유사한 구조 부재로 만들어졌지만, 동체보다 받는 스트레스가 적기 때문에 보통 더 가볍게 제작된다.

The other components of the typical empennage are of heavier construction than the tail cone.

전형적인 엠퍼너지의 다른 구성 요소는 꼬리 콘보다 더 무겁게 제작된다.

These members include fixed surfaces that help stabilize the aircraft and movable surfaces that help to direct an aircraft during flight.

이 구성 요소에는 항공기를 안정시키는 데 도움을 주는 고정된 표면과 비행 중 항공기를 조종하는 데 도움을 주는 가동식 표면이 포함된다.

The fixed surfaces are the horizontal stabilizer and vertical stabilizer.

고정된 표면에는 수평 안정판과 수직 안정판이 있다.

The movable surfaces are usually a rudder located at the aft edge of the vertical stabilizer and an elevator located at the aft edge of the horizontal stabilizer.

가동식 표면은 보통 수직 안정판의 후미 끝에 위치한 러더와 수평 안정판의 후미 끝에 위치한 엘리베이터로 구성된다.

4) Location Numbering Systems

The applicable manufacturer's numbering system and abbreviated designations or symbols should always be reviewed before attempting to locate a structural member.

해당 제조사의 번호 체계와 약어 또는 기호는 구조 부재의 위치를 찾기 전에 반드시 검토해야 한다.

They are not always the same.

이들은 항상 동일하지는 않다.

The following list includes location designations typical of those used by many manufacturers.

다음 목록은 많은 제조사들이 사용하는 위치 명칭의 일반적인 예를 포함한다.

• Fuselage stations (Fus. Sta. or FS) are numbered in inches from a reference or zero point known as the reference datum.

• 동체 기준점(Fus. Sta. 또는 FS)은 기준점(reference datum) 또는 제로 포인트에서 시작하여 인치 단위로 번호가 매겨진다.

• Buttock line or butt line (BL) is a vertical reference plane down the center of the aircraft from which measurements left or right can be made.

• 버턱 라인 또는 버트 라인(BL)은 항공기 중심선을 따라 내려가는 수직 기준 평면으로, 이 평면을 기준으로 좌우 측정을 할 수 있다.

• Water line (WL) is the measurement of height in inches perpendicular from a horizontal plane usually located at the ground, cabin floor, or some other easily referenced location.

• 워터 라인(WL)은 일반적으로 지면, 객실 바닥, 또는 기타 쉽게 참조할 수 있는 위치에 있는 수평면에서부터 인치 단위로 측정한 높이이다.

1-3. Flight Control Surfaces

1) Description

The directional control of a fixed-wing aircraft takes place around the lateral, longitudinal, and vertical axes by means of flight control surfaces designed to create movement about these axes.

고정익 항공기의 방향 제어는 비행 조종 면을 통해 횡축, 종축, 수직축을 중심으로 움직임을 생성함으로써 이루어진다.

These control devices are hinged or movable surfaces through which the attitude of an aircraft is controlled during takeoff, flight, and landing.

이 조종 장치는 이착륙 및 비행 중 항공기의 자세를 제어하는 힌지 또는 움직일 수 있는 면이다.

They are usually divided into two major

groups: primary or main flight control surfaces and secondary or auxiliary control surfaces.

이들은 보통 두 가지 주요 그룹으로 나뉜다: 주 조종 면 또는 주요 비행 조종 면과 보조 또는 부가 조종 면.

The primary flight control surfaces on a fixed-wing aircraft include: ailerons, elevators, and the rudder.

고정익 항공기의 주요 비행 조종 면에는 에일러론, 승강타, 방향타가 포함된다.

The ailerons are attached to the trailing edge of both wings and when moved, rotate the aircraft around the longitudinal axis.

에일러론은 양쪽 날개의 뒷전에 부착되어 있으며 움직일 때 항공기를 종축을 중심으로 회전시킨다.

The elevator is attached to the trailing edge of the horizontal stabilizer. When it is moved, it alters aircraft pitch, which is the attitude about the horizontal or lateral axis.

승강타는 수평 안정판의 뒷전에 부착되어 있다. 승강타가 움직이면 항공기의 자세가 횡축 또는 수평축을 중심으로 변화한다.

The rudder is hinged to the trailing edge of the vertical stabilizer. When the rudder changes position, the aircraft rotates about the vertical axis (yaw).

방향타는 수직 안정판의 뒷전에 힌지로 연결되어 있다. 방향타가 위치를 변경하면 항공기는 수직축을 중심으로 회전(요)한다.

Figure 1-7 shows the primary flight controls of a light aircraft and the movement they create relative to the three axes of flight.

그림 1-7은 경항공기의 주요 비행 조종 면과 비행의 세 축에 따른 움직임을 보여준다.

2) Flaps

Flaps are found on most aircraft.

플랩은 대부분의 항공기에 있다.

They are usually inboard on the wings' trailing edges adjacent to the fuselage.

플랩은 보통 날개의 후방 모서리(뒷전), 즉 동체와 인접한 곳에 위치한다.

Leading edge flaps are also common.

앞전 플랩도 일반적이다.

They extend forward and down from the inboard wing leading edge.

이들은 날개의 내부 앞전에서 앞으로 내려간다.

The flaps are lowered to increase the camber of the wings and provide greater lift and control at slow speeds.

플랩은 날개의 캠버(윙의 곡률)를 증가시키고, 저속에서 더 큰 양력과 조종을 제공하기 위해 내려진다.

They enable landing at slower speeds and shorten the amount of runway required for takeoff and landing.

플랩은 더 낮은 속도로 착륙할 수 있게 하며, 이륙과 착륙에 필요한 활주로의 길이를 줄여준다.

Flaps are usually constructed of materials and with techniques used on the other airfoils and control surfaces of a particular aircraft.

플랩은 일반적으로 해당 항공기의 다른 공기역학적 표면과 제어 표면에서 사용되는 재료와 기술로 제작된다.

Aluminum skin and structure flaps are the norm on light aircraft.

알루미늄 외피와 구조 플랩은 가벼운 항공기에서 일반 적이다.

Heavy and high-performance aircraft flaps may also be aluminum, but the use of composite structures is also common.

무거운 항공기나 고성능 항공기의 플랩도 알루미늄일 수 있지만, 복합 재료 구조를 사용하는 경우도 많다.

Chapter 2
Aerodynamics

2-1. General

1) Bernoulli's Principle

Bernoulli's principle states that when a fluid (air) flowing through a tube reaches a constriction, or narrowing, of the tube, the speed of the fluid flowing through that constriction is increased and its pressure is decreased.

베르누이의 원리에 따르면, 튜브를 흐르는 유체(공기)가 튜브의 좁아진 부분에 도달하면, 그 좁아진 부분을 흐르는 유체의 속도가 증가하고 압력은 감소한다고 한다.

The cambered (curved) surface of an airfoil (wing) affects the airflow exactly as a constriction in a tube affects airflow.

에어포일(날개)의 곡선(구부러진) 표면은 튜브에서의 좁아진 부분이 공기 흐름에 영향을 미치는 것처럼 공기 흐름에 영향을 미친다.

Diagram A of Figure 2-1 illustrates the effect of air passing through a constriction in a tube.

그림 2-1의 다이어그램 A는 튜브의 좁아진 부분을 통과하는 공기의 영향을 보여준다.

In Diagram B, air is flowing past a cambered surface, such as an airfoil, and the effect is similar to that of air passing through a restriction.

다이어그램 B에서는 공기가 에어포일과 같은 곡선 표면을 지나가고 있으며, 그 효과는 공기가 좁아진 부분을 통과하는 것과 비슷하다.

As the air flows over the upper surface of an airfoil, its velocity increases and its pressure decreases; an area of low pressure is formed.

공기가 에어포일의 윗면을 지나갈 때, 그 속도는 증가하고 압력은 감소하며, 저압 영역이 형성된다.

There is an area of greater pressure on the lower surface of the airfoil, and this greater pressure tends to move the wing upward.

에어포일의 아랫면에는 더 높은 압력 영역이 있으며, 이 높은 압력은 날개를 위쪽으로 밀어내는 경향이 있다.

The difference in pressure between the upper and lower surfaces of the wing is called lift.

날개 윗면과 아랫면 사이의 압력 차이는 양력(lift)이라고 한다.

2) Airfoil

(1) Description

An airfoil is any device that creates a force, based on Bernoulli's principles or Newton's laws, when air is caused to flow over the surface of the device.

에어포일은 공기가 장치의 표면 위로 흐르게 될 때, 베르누이의 원리나 뉴턴의 법칙을 기반으로 힘을 생성하는 장치이다.

An airfoil can be the wing of an airplane, the blade of a propeller, the rotor blade of a helicopter, or the fan blade of a turbo-fan engine.

에어포일은 항공기의 날개, 프로펠러의 날, 헬리콥터의 로터 날개, 또는 터보팬 엔진의 팬 날개일 수 있다.

The wing of an airplane moves through the air because the airplane is in motion, and generates lift by the process previously described.

항공기의 날개는 항공기가 움직이기 때문에 공기 중을 지나가며, 이전에 설명한 과정에 의해 양력을 생성한다.

In Figure 2-2 an airfoil, or wing, is shown, with some of the terminology that is used to describe a wing.

그림 2-2에서는 날개 또는 에어포일이 표시되어 있으며, 날개를 설명하는 데 사용되는 일부 용어들이 함께 나와 있다.

2-2. Forces in action during flight

In all types of flying, flight calculations are based on the magnitude and direction of four forces: weight, lift, drag, and thrust.

모든 종류의 비행에서 비행 계산은 네 가지 힘의 크기와 방향을 바탕으로 한다: 무게, 양력, 항력, 추진력.

An aircraft in flight is acted upon by four forces:

비행 중인 항공기는 네 가지 힘의 영향을 받는다:

1) Gravity or weight

The force that pulls the aircraft toward the earth.

항공기를 지구 쪽으로 끌어당기는 힘.

Weight is the force of gravity acting downward upon everything that goes into the aircraft, such as the aircraft itself, crew, fuel, and cargo.

무게는 중력이 항공기, 승무원, 연료, 화물 등 항공기 내 모든 것에 작용하여 아래로 끌어당기는 힘이다.

2) Lift

The force that pushes the aircraft upward.

항공기를 위로 밀어 올리는 힘.

Lift acts vertically and counteracts the effects of weight.

양력은 수직 방향으로 작용하며 무게의 영향을 상쇄시킨다.

3) Thrust

The force that moves the aircraft forward.

항공기를 앞으로 이동시키는 힘.

Thrust is the forward force produced by the powerplant that overcomes the force of drag.

추진력은 파워플랜트에 의해 생성된 전진하는 힘으로 항력의 힘을 이겨낸다.

4) Drag

The force that exerts a braking action to hold the aircraft back.

항공기를 뒤로 제지하는 제동 작용을 일으키는 힘.

Drag is a backward deterrent force and is caused by the disruption of the airflow by the wings, fuselage, and protruding objects.

항력은 뒤로 향하는 방해하는 힘이며 날개, 동체, 돌출된 물체들에 의해 공기 흐름이 방해되어 발생한다.

2-3. The Axes of an Aircraft

Whenever an aircraft changes its attitude in flight, it must turn about one or more of three axes.

항공기가 비행 중에 자세를 변경할 때, 그것은 세 가지 축 중 하나 또는 그 이상의 축을 중심으로 회전해야 한다.

Figure 2-4 shows the three axes, which are imaginary lines passing through the center of the aircraft.

그림 2-4는 항공기 중심을 통과하는 가상의 세 축을 보여준다.

The axes of an aircraft can be considered as imaginary axles around which the aircraft turns like a wheel.

항공기의 축은 항공기가 바퀴처럼 회전하는 가상의 축으로 간주할 수 있다.

At the center, where all three axes intersect, each is perpendicular to the other two.

모든 세 축이 교차하는 중심에서는 각 축이 나머지 두 축에 수직이다.

The axis that extends lengthwise through the fuselage from the nose to the tail is called the longitudinal axis.

동체를 따라 앞쪽에서 뒤쪽까지 길게 뻗어 있는 축은 종축(longitudinal axis)이라고 한다.

The axis that extends crosswise from wing tip to wing tip is the lateral, or pitch, axis.

날개 끝에서 날개 끝까지 가로로 뻗어 있는 축은 측면축(lateral axis) 또는 피치 축(pitch axis)이라고 한다.

The axis that passes through the center, from top to bottom, is called the vertical, or yaw, axis.

위에서 아래로 지나가는 축은 수직축(vertical axis) 또는 요축(yaw axis)이라고 한다.

Roll, pitch, and yaw are controlled by three control surfaces.

롤, 피치, 요는 세 가지 조종면에 의해 제어된다.

Roll is produced by the ailerons, which are located at the trailing edges of the wings.

롤은 날개의 후방 가장자리에 위치한 에일러론에 의해 발생한다.

Pitch is affected by the elevators, the rear portion of the horizontal tail assembly.

피치는 수평 꼬리날개의 후방 부분에 위치한 엘리베이터에 의해 영향을 받는다.

Yaw is controlled by the rudder, the rear portion of the vertical tail assembly.

요는 수직 꼬리날개의 후방 부분에 위치한 러더에 의해 제어된다.

Chapter 3
Advanced Composite Materials

3-1. General

Composite materials are becoming more important in the construction of aerospace structures.

복합 재료는 항공 우주 구조물 제작에서 점점 더 중요해지고 있다.

Aircraft parts made from composite materials, such as fairings, spoilers, and flight controls, were developed during the 1960s for their weight savings over aluminum parts.

복합 재료로 만든 항공기 부품, 예를 들면 페어링, 스포일러, 비행 조종 장치는 알루미늄 부품보다 무게 절감을 위해 1960년대에 개발되었다.

New generation large aircraft are designed with all composite fuselage and wing structures, and the repair of these advanced composite materials requires an in-depth knowledge of composite structures, materials, and tooling.

신세대 대형 항공기는 전체 복합재 동체와 날개 구조로 설계되며, 이러한 고급 복합 재료의 수리는 복합 재료 구조, 재료, 치공구에 대한 심층적인 지식이 필요하다.

The primary advantages of composite materials are their high strength, relatively low weight, and corrosion resistance.

복합 재료의 주요 장점은 높은 강도, 상대적으로 낮은 무게, 그리고 내식성이다.

3-2. Composite Material and Process

1) Prepreg

Prepreg is a fabric or tape that is impregnated with a resin during the manufacturing process.

프리프레그는 제조 과정 중에 수지로 침투된 직물 또는 테이프이다.

The resin system is already mixed and is in the B stage cure.

수지 시스템은 이미 혼합되어 있고 B단계 경화 상태에 있다.

Store the prepreg material in a freezer below 0 °F to prevent further curing of the resin.

프리프레그 재료는 수지의 추가 경화를 방지하기 위해 0°F 이하의 냉동고에 보관해야 한다.

You must remove the prepreg from the freezer and let the material thaw, which might take 8 hours for a full roll.

프리프레그는 냉동고에서 꺼내어 재료가 해동되도록 해야 하며, 전체 롤의 경우 해동에 8시간 정도 걸릴 수 있다.

Store the prepreg materials in a sealed, moisture proof bag.

프리프레그 재료는 밀봉된 방습 백에 보관해야 한다.

Do not open these bags until the material is completely thawed, to prevent contamination of the material by moisture.

이 백은 재료가 완전히 해동될 때까지 열지 말아야 하며, 이는 재료가 수분에 의해 오염되는 것을 방지하기 위함이다.

2) Vacuum Bag

The vacuum bag material provides a tough layer between the repair and the atmosphere.

진공 백 재료는 수리와 대기 사이에 강한 층을 제공한다.

The vacuum bag material is available in different temperature ratings, so make sure that the material used for the repair can handle the cure temperature.

진공 백 재료는 다양한 온도 등급으로 제공되므로 수리를 위해 사용되는 재료가 경화 온도를 처리할 수 있는지 확인해야 한다.

Most vacuum bag materials are one time use, but material made from flexible silicon rubber is reusable.

대부분의 진공 백 재료는 한 번만 사용되지만, 유연한 실리콘 고무로 만들어진 재료는 재사용이 가능하다.

3-3. Composite Cure Equipment

1) Autoclave

An autoclave system allows a complex chemical reaction to occur inside a pressure vessel according to a specified time, temperature, and pressure profile in order to process a variety of materials.

오토클레이브 시스템은 지정된 시간, 온도, 압력 프로파일에 따라 다양한 재료를 처리하기 위해 압력 용기 내부에서 복잡한 화학 반응이 일어나도록 허용한다.

Autoclaves that are operated at lower temperatures and pressures can be pressurized by air, but if higher temperatures and pressures are required for the cure cycle, a 50/50 mixture of air and nitrogen or 100 percent nitrogen should be used to reduce the chance of an autoclave fire.

낮은 온도와 압력에서 작동하는 오토클레이브는 공기로 압력을 가할 수 있지만, 경화 사이클에 더 높은 온도와 압력이 필요하다면, 오토클레이브 화재의 위험을 줄이기 위해 공기와 질소의 50/50 혼합물 또는 100% 질소를 사용해야 한다.

The major elements of an autoclave system are a vessel to contain pressure, sources to heat the gas stream and circulate it uniformly within the vessel, a subsystem to apply vacuum to parts covered by a vacuum bag, a subsystem to control operating parameters, and a subsystem to load the molds into the autoclave.

오토클레이브 시스템의 주요 구성 요소는 압력을 담을 수 있는 용기, 가스 흐름을 가열하고 용기 내에서 균일하게 순환시키는 열원, 진공백으로 덮인 부품에 진공을 적용하는 서브시스템, 운영 매개변수를 제어하는 서브시스템, 오토클레이브에 몰드를 장입하는 서브시스템이다.

Modern autoclaves are computer controlled and the operator can write and monitor all types of cure cycle programs.

현대적인 오토클레이브는 컴퓨터로 제어되며, 운영자는 모든 유형의 경화 사이클 프로그램을 작성하고 모니터링할 수 있다.

The most accurate way to control the cure cycle is to control the autoclave controller with thermocouples that are placed on the actual part.

경화 사이클을 제어하는 가장 정확한 방법은 실제 부품에 배치된 열전대를 사용하여 오토클레이브 제어기를 제어하는 것이다.

2) Heat Bonder

A heat bonder is a portable device that automatically controls heating based on temperature feedback from the repair area.

히트 본더는 수리 구역에서 온도 피드백을 기반으로 가열을 자동으로 제어하는 휴대용 장치이다.

Heat bonders also have a vacuum pump that supplies and monitors the vacuum in the vacuum bag.

히트 본더에는 또한 진공 팽창백 내의 진공을 공급하고 모니터링하는 진공 펌프가 있다.

Heat Blanket

A heat blanket is a flexible heater.

히트 블랭킷은 유연한 히터이다.

It is made of two layers of silicon rubber

with a metal resistance heater between the two layers of silicon.

이것은 두 층의 실리콘 고무와 그 사이에 금속 저항 히터가 있는 구조로 만들어진다.

Heat blankets are a common method of applying heat for repairs on the aircraft.

히트 블랭킷은 항공기 수리를 위한 열 적용 방법 중 하나로 일반적으로 사용된다.

3-4. Inspection

1) Ultrasonic Bond Tester Inspection

Low-frequency and high-frequency bond testers are used for ultrasonic inspections of composite structures.

저주파 및 고주파 본드 시험기는 복합 재료 구조의 초음파 검사에 사용된다.

These bond testers use an inspection probe that has one or two transducers.

이 본드 시험기는 하나 또는 두 개의 변환기가 있는 검사 탐침을 사용한다.

The high-frequency bond tester is used to detect delaminations and voids.

고주파 본드 시험기는 박리와 공극을 검출하는 데 사용된다.

It cannot detect a skin-to-honeycomb core disbond or porosity.

이 시험기는 외피와 벌집 코어의 접착력 상실이나 기공을 검출할 수 없다.

It can detect defects as small as 0.5-inch in diameter.

이 시험기는 지름이 0.5인치만큼 작은 결함도 검출할 수 있다.

The low-frequency bond tester uses two transducers and is used to detect delamination, voids, and skin to honeycomb core disbands.

저주파 본드 시험기는 두 개의 변환기를 사용하며, 박리, 공간, 외피와 벌집 코어의 접착력 상실을 검출하는 데 사용된다.

This inspection method does not detect which side of the part is damaged, and cannot detect defects smaller than 1.0-inch.

이 검사 방법은 부품의 어느 쪽이 손상되었는지 감지하지 않으며, 1.0인치보다 작은 결함은 검출할 수 없다.

2) Audible Sonic Testing (Coin Tapping)

Sometimes referred to as audio, sonic, or coin tap, this technique makes use of frequencies in the audible range (10 Hz to 20 Hz).

때때로 오디오, 음파, 또는 동전 두드리기라고 불리는 이 기술은 가청 범위(10 Hz에서 20 Hz) 내의 주파수를 사용한다.

A surprisingly accurate method in the hands of experienced personnel, tap testing is perhaps the most common technique used for the detection of delamination and/or disbond.

경험이 풍부한 인력에게는 놀랍도록 정확한 방법으로, 탭 테스트는 아마도 박리 및/또는 탈착 검출에 가장 일반적으로 사용되는 기술일 것이다.

Chapter 4

Aircraft Systems

4-1. Aircraft Instrument Systems

1) Introduction

Since the beginning of manned flight, it has been recognized that supplying the pilot with information about the aircraft and its operation could be useful and lead to safer flight.

유인 비행이 시작된 이후, 파일럿에게 항공기와 그 작동에 대한 정보를 제공하는 것이 유용하고 더 안전한 비행으로 이어질 수 있다는 것이 인식되었다.

The Wright Brothers had very few instruments on their Wright Flyer, but they did have an engine tachometer, an anemometer (wind meter), and a stop watch.

라이트 형제는 그들의 라이트 플라이어에 매우 적은 수의 계기를 장착했지만, 엔진 회전계, 풍속계(바람 측정기), 그리고 스톱워치(시계)를 가지고 있었다.

They were obviously concerned about the aircraft's engine and the progress of their flight.

그들은 명백히 항공기의 엔진과 비행 진행 상황에 대해 걱정하고 있었다.

From that simple beginning, a wide variety of instruments have been developed to inform flight crews of different parameters.

그 간단한 시작에서, 비행 승무원들에게 다양한 매개변수에 대한 정보를 제공하기 위해 여러 가지 계기들이 개발되었다.

Instrument systems now exist to provide information on the condition of the aircraft, engine, components, the aircraft's attitude in the sky, weather, cabin environment, navigation, and communication.

현재 계기 시스템은 항공기, 엔진, 구성 요소, 항공기의 자세, 날씨, 객실 환경, 항법 및 통신에 대한 정보를 제공하는 시스템이 존재한다.

2) Classifying Instruments

There are three basic kinds of instruments classified by the job they perform: flight instruments, engine instruments, and navigation instruments.

기본적으로 수행하는 작업에 따라 세 가지 종류의 계기가 있다: 비행 계기, 엔진 계기, 항법 계기.

There are also miscellaneous gauges and indicators that provide information that do not fall into these classifications, especially on large complex aircraft.

또한, 이러한 분류에 속하지 않는 정보를 제공하는 다양한 게이지와 지시기들이 있다. 특히 대형 복잡한 항공기에서 그렇다.

Flight control position, cabin environmental systems, electrical power, and auxiliary power units (APUs), for example, are all monitored and controlled from the cockpit via the use of instruments systems.

예를 들어, 비행 제어 위치, 객실 환경 시스템, 전기력, 보조 동력 장치(APU) 등은 모두 계기 시스템을 사용하여 조종석에서 모니터링하고 제어된다.

(1) Flight Instruments

The instruments used in controlling the aircraft's flight attitude are known as the flight instruments.

항공기의 비행 자세를 제어하는 데 사용되는 기기는 비행 계기라고 알려져 있다.

There are basic flight instruments, such as the altimeter that displays aircraft altitude; the airspeed indicator; and the magnetic direction indicator, a form of compass.

기본적인 비행 계기로는 항공기의 고도를 표시하는 고도계, 속도계, 그리고 자석 방향 표시기(나침반의 일종)가 있다.

Additionally, an artificial horizon, turn coordinator, and vertical speed indicator are flight instruments present in most aircraft.

또한, 인공 수평선 계기, 선회계(회전 조정기), 수직 속도 표시기 등이 대부분의 항공기에 있는 비행 계기이다.

Over the years, flight instruments have come to be situated similarly on the instrument panels in most aircraft.

수년 동안, 비행 기기들은 대부분의 항공기에서 계기판에 비슷하게 배치되게 되었다.

This basic T arrangement for flight instruments is shown in Figure 4-2.

비행 계기를 위한 이 기본적인 T자 형태 배치는 그림 4-2에서 볼 수 있다.

The top center position directly in front of the pilot and copilot is the basic display position for the artificial horizon even in modern glass cockpits (those with solid-state, flat-panel screen indicating systems).

조종사와 부조종사 바로 앞에 있는 상단 중앙 위치는 현대식 유리 조종석(고체 상태의 평면 화면 표시 시스템이 있는 조종석)에서도 인공 수평선의 기본 표시 위치이다.

(2) Engine Instruments

Engine instruments are those designed to measure operating parameters of the aircraft's engine.

엔진 계기는 항공기 엔진의 작동 파라미터를 측정하도록 설계된 것이다.

These are usually quantity, pressure, and temperature indications.

이들은 보통 양, 압력 및 온도 표시가 포함된다.

They also include measuring engine speed.

또한 엔진 속도를 측정하는 것도 포함된다.

The most common engine instruments are the fuel and oil quantity and pressure gauges, tachometers, and temperature gauges.

가장 일반적인 엔진 계기로는 연료 및 오일 양과 압력 게이지, 타코미터(엔진 회전수 계기), 온도 게이지가 있다.

Engine instrumentation is often displayed in the center of the cockpit where it is easily visible to the pilot and copilot.

엔진 계기는 종종 조종사와 부조종사가 쉽게 볼 수 있는 조종석 중앙에 표시된다.

On light aircraft requiring only one flight crew member, this may not be the case.

단일 비행 승무원이 필요한 경비행기의 경우, 이것이 해당되지 않을 수 있다.

Multiengine aircraft often use a single gauge for a particular engine parameter, but it displays information for all engines through the use of multiple pointers on the same dial face.

다발 엔진 항공기는 특정 엔진 파라미터에 대해 단일 게이지를 사용하지만, 동일한 다이얼 면에 여러 개의 지시기를 사용하여 모든 엔진의 정보를 표시한다.

(3) Navigation Instruments

Navigation instruments are those that contribute information used by the pilot to guide the aircraft along a definite course.

항법 계기는 조종사가 항공기를 정해진 경로로 안내하는 데 사용하는 정보를 제공하는 기기들이다.

This group includes compasses of various kinds, some of which incorporate the use of radio signals to define a specific course while flying the aircraft en route from one airport to another.

이 그룹에는 여러 종류의 나침반이 포함되어 있으며, 그 중 일부는 항공기가 한 공항에서 다른 공항으로 이동할 때 특정 경로를 정의하기 위해 라디오 신호를 사용한다.

Other navigational instruments are designed specifically to direct the pilot's approach to landing at an airport.

다른 항법 계기들은 조종사가 공항에 착륙하는 접근을 안내하는 데 특별히 설계된다.

Traditional navigation instruments include a clock and a magnetic compass.

전통적인 항법 계기에는 시계와 자석 나침반이 포함된다.

Along with the airspeed indicator and wind information, these can be used to calculate navigational progress.

속도계와 풍속 정보와 함께, 이것들은 항법 진행 상황을 계산하는 데 사용될 수 있다.

Radios and instruments sending locating information via radio waves have replaced these manual efforts in modern aircraft.

라디오와 라디오 파장을 통해 위치 정보를 전송하는 계기들은 현대 항공기에서 이러한 수동적인 방법들을 대체했다.

Global position systems (GPS) use satellites to pinpoint the location of the aircraft via geometric triangulation.

위치 정보 시스템(GPS)은 위성을 사용하여 기하학적 삼각측량을 통해 항공기의 위치를 정확하게 찾는다.

This technology is built into some aircraft instrument packages for navigational purposes.

이 기술은 일부 항공기 계기 패키지에 항법 목적을 위해 내장되어 있다.

4-2. Aircraft Hydraulic Systems

1) Introduction

Hydraulic systems are not new to aviation.

유압 시스템은 항공에 새로운 기술이 아니다.

Early aircraft had hydraulic brake systems.

초기 항공기들은 유압 제동 시스템을 가졌다.

Hydraulic systems in aircraft provide a means for the operation of aircraft components.

항공기의 유압 시스템은 항공기 구성 요소의 작동을 위한 수단을 제공한다.

The operation of landing gear, flaps, flight control surfaces, and brakes is largely accomplished with hydraulic power systems.

착륙 장치, 플랩, 비행 제어 표면, 그리고 제동 장치의 작동은 주로 유압 시스템으로 이루어진다.

Hydraulic system complexity varies from small aircraft that require fluid only for manual operation of the wheel brakes to large transport aircraft where the systems are large and complex.

유압 시스템의 복잡성은 바퀴 제동을 수동으로 조작하는 데만 유체가 필요한 소형 항공기에서 시스템이 크고 복잡한 대형 수송 항공기까지 다양하다.

To achieve the necessary redundancy and reliability, the system may consist of several subsystems.

필요한 이중화와 신뢰성을 달성하기 위해, 시스템은 여러 개의 하위 시스템으로 구성될 수 있다.

Each subsystem has a power generating device (pump), reservoir, accumulator, heat exchanger, filtering system, etc.

각 하위 시스템은 전력 생성 장치(펌프), 저장소, 축전기, 열 교환기, 필터링 시스템 등을 갖추고 있다.

System operating pressure may vary from a couple hundred pounds per square inch (psi) in small aircraft and rotorcraft to 5,000 psi in large transports.

시스템 작동 압력은 소형 항공기와 회전익 항공기의 경우 psi에서 대형 수송 항공기에서는 5,000 psi까지 다양할 수 있다.

2) Hydraulic System Components

(1) Reservoirs

The reservoir is a tank in which an adequate supply of fluid for the system is stored.

저장소는 시스템을 위한 충분한 양의 액체가 저장되는 탱크이다.

Fluid flows from the reservoir to the pump, where it is forced through the system and eventually returned to the reservoir.

액체는 저장소에서 펌프로 흐르고, 그곳에서 시스템을 통해 강제로 흐르며 결국 저장소로 돌아온다.

The reservoir not only supplies the operating needs of the system, but it also replenishes fluid lost through leakage.

저장소는 시스템의 운영에 필요한 액체를 공급할 뿐만 아니라, 누수로 잃어버린 액체를 보충하기도 한다.

Furthermore, the reservoir serves as an overflow basin for excess fluid forced out of the system by thermal expansion (the increase of fluid volume caused by temperature changes), the accumulators, and by piston and rod displacement.

게다가 저장소는 열 팽창(온도 변화로 인한 액체 부피 증가), 축압기, 그리고 피스톤과 로드 이동으로 인해 시스템에서 밀려 나오는 과잉 액체를 위한 오버플로우 저수지 역할도 한다.

The reservoir also furnishes a place for the fluid to purge itself of air bubbles that may enter the system.

저장소는 또한 시스템에 들어올 수 있는 공기 방울을 액체가 스스로 제거할 수 있는 장소를 제공한다.

Foreign matter picked up in the system may also be separated from the fluid in the reservoir or as it flows through line filters.

시스템에서 집수된 이물질은 저장소에서 액체와 분리되거나 액체가 라인 필터를 통과할 때 분리될 수 있다.

Reservoirs are either pressurized or nonpressurized.

저장소는 압력이 있거나 없을 수 있다.

(2) Check Valve

A check valve allows fluid to flow unimpeded in one direction but prevents or restricts fluid flow in the opposite direction.

체크 밸브는 한 방향으로는 유체가 방해받지 않고 흐르게 하며, 반대 방향으로는 유체 흐름을 방지하거나 제한한다.

(3) Hydraulic Fuses

A hydraulic fuse is a safety device.

유압 퓨즈는 안전 장치이다.

Fuses may be installed at strategic locations throughout a hydraulic system.

퓨즈는 유압 시스템의 전략적인 위치에 설치될 수 있다.

They detect a sudden increase in flow, such as a burst downstream, and shut off the fluid flow.

퓨즈는 흐름의 갑작스러운 증가, 하류에서의 폭발,를 감지하고 유체 흐름을 차단한다.

(4) Relief Valves

Hydraulic pressure must be regulated in order to use it to perform the desired tasks.

유압 압력은 원하는 작업을 수행하기 위해 조절되어야 한다.

A pressure relief valve is used to limit the amount of pressure being exerted on a confined liquid.

압력 릴리프 밸브는 제한된 액체에 가해지는 압력의 양을 제한하는 데 사용된다.

This is necessary to prevent failure of components or rupture of hydraulic lines under excessive pressures.

이는 과도한 압력으로 인한 부품 고장이나 유압 라인의 파열을 방지하기 위해 필요하다.

The pressure relief valve is, in effect, a system safety valve.

압력 릴리프 밸브는 사실상 시스템 안전 밸브이다.

(5) Actuators

An actuating cylinder transforms energy in the form of fluid pressure into mechanical force, or action, to perform work.

구동 실린더는 유체 압력 형태의 에너지를 기계적인 힘 또는 동작으로 변환하여 작업을 수행한다.

It is used to impart powered linear motion to some movable object or mechanism.

이는 일부 이동 가능한 물체나 메커니즘에 동력화된 직선 운동을 전달하는 데 사용된다.

Actuating cylinders are of two major types: single action and double action.

구동 실린더는 두 가지 주요 유형이 있다: 단일 작용과 이중 작용.

The single-action (single port) actuating cylinder is capable of producing powered movement in one direction only.

단일 작용(단일 포트) 구동 실린더는 한 방향으로만 동력 운동을 생성할 수 있다.

The double-action (two ports) actuating cylinder is capable of producing powered movement in two directions.

이중 작용(두 포트) 구동 실린더는 두 방향으로 동력 운동을 생성할 수 있다.

4-3. Aircraft Pneumatic Systems

1) Introduction

Some aircraft manufacturers have equipped their aircraft with a high pressure pneumatic system (3,000 psi) in the past.

일부 항공기 제조업체들은 과거에 고압 공기압 시스템 (3,000 psi)을 항공기에 장착했다.

The last aircraft to utilize this type of system was the Fokker F27.

이러한 시스템을 사용한 마지막 항공기는 Fokker F27 이었다.

Such systems operate a great deal like hydraulic systems, except they employ air instead of a liquid for transmitting power.

이러한 시스템은 유압 시스템처럼 작동하지만, 파워를 전달할 때 액체 대신 공기를 사용한다.

Pneumatic systems are sometimes used for:

공기압 시스템은 때때로 다음 용도로 사용된다:

• Brakes
• 브레이크

• Opening and closing doors
• 문 열기와 닫기

• Driving hydraulic pumps, alternators, starters, water injection pumps, etc.
• 유압 펌프, 발전기, 스타터, 수분 주입 펌프 등 구동

• Operating emergency devices
• 비상 장치 작동

2) Pneumatic System Components

Pneumatic systems are often compared to hydraulic systems.

공기압 시스템은 종종 유압 시스템과 비교된다.

Pneumatic systems do not utilize reservoirs, hand pumps, accumulators, regulators, or engine-driven or electrically driven power pumps for building normal pressure.

공기압 시스템은 정상 압력을 생성하기 위해 저장소, 수동 펌프, 축압기, 조절기, 엔진 구동 또는 전기 구동 전력 펌프를 사용하지 않는다.

But similarities do exist in some components.

하지만 일부 부품에서는 유사성이 존재한다.

(1) Air Compressors

On some aircraft, permanently installed air compressors have been added to recharge air bottles whenever pressure is used for operating a unit.

일부 항공기에서는 압력이 장치를 작동시키는 데 사용될 때마다 공기통을 재충전하기 위해 영구적으로 설치된 공기 압축기를 추가했다.

Several types of compressors are used for this purpose.

이 목적을 위해 여러 종류의 압축기가 사용된다.

Some have two stages of compression, while others have three, depending on the maximum desired operating pressure.

일부는 두 단계 압축을 가지고 있고, 다른 일부는 최대 원하는 작동 압력에 따라 세 단계를 가진다.

(2) Relief Valves

Relief valves are used in pneumatic systems to prevent damage.

릴리프 밸브는 손상을 방지하기 위해 공기압 시스템에서 사용된다.

They act as pressure limiting units and prevent excessive pressures from bursting lines and blowing out seals.

이들은 압력 제한 장치로 작용하며, 과도한 압력이 라인을 터뜨리거나 씰을 터뜨리는 것을 방지한다.

(3) Check Valves

Check valves are used in both hydraulic and pneumatic systems.

체크 밸브는 유압 시스템과 공압 시스템 모두에서 사용된다.

Figure 4-11 illustrates a flap-type pneumatic check valve.

그림 4-11은 플랩 타입 공압 체크 밸브를 보여준다.

Air enters the left port of the check valve, compresses a light spring, forcing the check valve open and allowing air to flow out the right port.

공기는 체크 밸브의 왼쪽 포트를 통해 들어가서 가벼운 스프링을 압축시키고, 체크 밸브를 열어 오른쪽 포트를 통해 공기가 흐를 수 있도록 한다.

But if air enters from the right, air pressure closes the valve, preventing a flow of air out the left port.

하지만 공기가 오른쪽에서 들어오면, 공기 압력이 밸브를 닫아서 왼쪽 포트로 공기가 흐르는 것을 막는다.

Thus, a pneumatic check valve is a one-direction flow control valve.

따라서, 공압 체크 밸브는 일방향 흐름 제어 밸브이다.

(4) Restrictors

Restrictors are a type of control valve used in pneumatic systems.

흐름제한장치는 공압 시스템에서 사용되는 제어 밸브의 일종이다.

Figure 4-12 illustrates an orifice-type restrictor with a large inlet port and a small outlet port.

그림 4-12는 큰 입구 포트와 작은 출구 포트를 가진 오리피스형 흐름제한장치를 보여준다.

The small outlet port reduces the rate of airflow and the speed of operation of an actuating unit.

작은 출구 포트는 공기 흐름의 속도와 작동 유닛의 속도를 감소시킨다.

(5) Filters

Pneumatic systems are protected against dirt by means of various types of filters.

공압 시스템은 다양한 종류의 필터를 사용하여 먼지로부터 보호된다.

A micronic filter consists of a housing with two ports, a replaceable cartridge, and a relief valve.

마이크로닉 필터는 두 개의 포트가 있는 하우징, 교체 가능한 카트리지, 그리고 릴리프 밸브로 구성된다.

4-4. Aircraft Landing Gear Systems

1) Introduction

Aircraft landing gear supports the entire weight of an aircraft during landing and ground operations.

항공기 랜딩 기어는 착륙 및 지상 작업 중 항공기의 전체 무게를 지지한다.

They are attached to primary structural members of the aircraft.

이들은 항공기의 주요 구조 구성 요소에 부착된다.

The type of gear depends on the aircraft design and its intended use.

기어의 종류는 항공기 설계와 그 용도에 따라 달라진다.

Most landing gear have wheels to facilitate operation to and from hard surfaces, such as airport runways.

대부분의 랜딩 기어는 공항 활주로와 같은 단단한 표면에서 오가며 작동을 용이하게 하기 위해 바퀴를 갖추고 있다.

Other gear feature skids for this purpose, such as those found on helicopters, balloon gondolas, and in the tail area of some taildragger aircraft.

다른 기어는 이 목적을 위해 스키드를 갖추고 있다. 예를 들어, 헬리콥터, 풍선 곤돌라 및 일부 테일드래거 항공기의 꼬리 부분에 해당한다.

Aircraft that operate to and from frozen lakes and snowy areas may be equipped with landing gear that have skis.

얼어붙은 호수와 눈 덮인 지역에서 운항하는 항공기는 스키가 장착된 랜딩 기어를 장착할 수 있다.

Aircraft that operate to and from the surface of water have pontoon-type landing gear.

수면에서 운항하는 항공기는 플로트 타입 랜딩 기어를 갖추고 있다.

Regardless of the type of landing gear utilized, shock absorbing equipment, brakes, retraction mechanisms, controls, warning devices, cowling, fairings, and structural members necessary to attach the gear to the aircraft are considered parts of the landing gear system.

사용되는 랜딩 기어의 종류에 관계없이, 충격 흡수 장비, 브레이크, 수축 기구, 제어 장치, 경고 장치, 카울링, 페어링 및 기어를 항공기에 부착하는 데 필요한 구조 구성 요소들은 랜딩 기어 시스템의 일부로 간주된다.

2) Landing Gear Arrangement

Three basic arrangements of landing gear are used: tail wheel-type landing gear (also known as conventional gear), tandem landing gear, and tricycle-type landing gear.

착륙 장치 배열에는 세 가지 기본 배열이 사용된다: 꼬리바퀴식 착륙 장치(전통적인 기어라고도 불린다), 직렬식 착륙 장치, 그리고 삼륜식 착륙 장치.

(1) Tail Wheel-Type Landing Gear

Tail wheel-type landing gear is also known as conventional gear because many early aircraft use this type of arrangement.

꼬리바퀴식 착륙 장치는 많은 초기 항공기들이 이 배열을 사용했기 때문에 전통적인 기어라고도 불린다.

The main gear are located forward of the center of gravity, causing the tail to require support from a third wheel assembly.

주 착륙 장치는 무게 중심의 앞쪽에 위치하고, 그로 인해 테일이 세 번째 바퀴 조립체의 지원을 필요로 한다.

(2) Tandem Landing Gear

Few aircraft are designed with tandem landing gear.

몇몇 항공기만이 직렬식 착륙 장치를 설계한다.

As the name implies, this type of landing gear has the main gear and tail gear aligned on the longitudinal axis of the aircraft.

이름이 암시하는 것처럼, 이 유형의 착륙 장치는 항공기의 종축(axis)에 맞춰 주 착륙 장치(main gear)와 꼬리 착륙 장치(tail gear)가 정렬된다.

Sailplanes commonly use tandem gear, although many only have one actual gear forward on the fuselage with a skid under the tail.

글라이더는 일반적으로 직렬식 착륙 장치를 사용하지만, 많은 글라이더는 실제 착륙 장치가 항공기 동체 앞부분에 하나만 있고 꼬리 부분에는 스키드(skid)가 있다.

(3) Tricycle-Type Landing Gear

The most commonly used landing gear arrangement is the tricycle-type landing gear.

가장 일반적으로 사용되는 착륙 장치 배열은 삼륜형 착륙 장치이다.

It is comprised of main gear and nose gear.

이는 주 착륙 장치(main gear)와 전방 착륙 장치(nose gear)로 구성된다.

The tricycle-type landing gear arrangement consists of many parts and assemblies.

삼륜형 착륙 장치 배열은 많은 부품과 조합으로 구성된다.

These include air/oil shock struts, gear alignment units, support units, retraction and safety devices, steering systems, wheel and brake assemblies, etc.

이들에는 공기/오일 충격 스트럿, 기어 정렬 장치, 지지 장치, 수축(접힘) 및 안전 장치, 조향 시스템, 바퀴 및 브레이크 조립체 등이 포함된다.

(4) Fixed and Retractable Landing Gear

Further classification of aircraft landing gear can be made into two categories: fixed and retractable.

항공기 착륙 장비는 고정형과 접힘(수축)형 두 가지 범주로 추가 분류할 수 있다.

Many small, single-engine light aircraft have fixed landing gear, as do a few light twins.

많은 소형 단발 경비행기와 몇몇 경량 쌍발 항공기는 고정형 착륙 장치를 가지고 있다.

This means the gear is attached to the airframe and remains exposed to the slipstream as the aircraft is flown.

이것은 착륙 장비가 기체에 부착되어 항공기가 비행하는 동안 슬립스트림에 노출된 상태로 유지된다는 것을 의미한다.

As the speed of an aircraft increases, so does parasite drag.

항공기의 속도가 증가함에 따라, 유해 항력도 증가한다.

Mechanisms to retract and stow the landing gear to eliminate parasite drag add weight to the aircraft.

유해 항력을 제거하기 위해 착륙 장치를 접고(올리고) 보관하는 기계장치가 항공기에 추가 중량을 더한다.

On slow aircraft, the penalty of this added weight is not overcome by the reduction of drag, so fixed gear is used.

속도가 느린 항공기에서는 이 추가 중량에 대한 패널티가 항력 감소로 보상되지 않기 때문에 고정형 착륙 장치가 사용된다.

As the speed of the aircraft increases, the drag caused by the landing gear becomes greater and a means to retract the gear to eliminate parasite drag is required, despite the weight of the mechanism.

항공기의 속도가 증가함에 따라 착륙 장치로 인한 항력이 커지고, 기계장치의 무게에도 불구하고 유해 항력을 제거하기 위해 착륙 장치를 접을 수 있는 방법이 필요하다.

3) Nose Wheel Steering Systems

The nose wheel on most aircraft is steerable from the flight deck via a nose wheel steering system.

대부분의 항공기에서 노즈 휠은 노즈 휠 조종 시스템을 통해 조종석에서 조향이 가능하다.

This allows the aircraft to be directed during ground operation.

이 시스템은 항공기가 지상 작동 중에 방향을 조정할 수 있게 한다.

(1) Small Aircraft

Most small aircraft have steering capabilities through the use of a simple system of mechanical linkages connected to the rudder pedals.

대부분의 소형 항공기는 방향타 페달에 연결된 간단한 기계적 연결 시스템을 사용하여 조향 능력을 가진다.

Push-pull tubes are connected to pedal horns on the lower strut cylinder.

푸시-풀 튜브는 하부 스트럿 실린더에 있는 페달 호른에 연결되어 있다.

As the pedals are depressed, the movement is transferred to the strut piston axle and wheel assembly which rotates to the left or right.

페달이 눌리면 그 움직임이 스트럿 피스톤 축과 휠 어셈블리에 전달되고, 이는 좌우로 회전한다.

(2) Large Aircraft

Due to their mass and the need for positive control, large aircraft utilize a power source for nose wheel steering.

대형 항공기는 그들의 질량과 정확한 조정을 위해 전방 휠 조향을 위한 동력원을 사용한다.

Hydraulic power predominates.

유압 동력이 주로 사용된다.

There are many different designs for large aircraft nose steering systems.

대형 항공기의 전방 휠 조향 시스템에는 다양한 설계가 있다.

Most share similar characteristics and components.

대부분은 비슷한 특성과 구성 요소를 공유한다.

Control of the steering is from the flight deck through the use of a small wheel, tiller, or joystick typically mounted on the left side wall.

조향의 제어는 비행 조종석에서 작은 휠, 조타기, 또는 조이스틱을 사용하여 이루어지며, 이는 일반적으로 왼쪽 벽에 장착된다.

4-5. Aircraft Fuel Systems

1) Types of Aviation Fuel

Each aircraft engine is designed to burn a

certain fuel.

각 항공기 엔진은 특정 연료를 연소하도록 설계되어 있다.

Use only the fuel specified by the manufacturer.

제조업체가 지정한 연료만 사용해야 한다.

Mixing fuels is not permitted.

연료 혼합은 허용되지 않는다.

There are two basic types of fuel discussed in this section: reciprocating-engine fuel (also known as gasoline or AVGAS) and turbine-engine fuel (also known as jet fuel or kerosene).

이 섹션에서는 두 가지 기본 연료 종류에 대해 다룬다: 왕복엔진 연료(가솔린 또는 AVGAS라고도 함)와 터빈 엔진 연료(제트 연료 또는 등유라고도 함).

(1) Reciprocating Engine Fuel

Reciprocating engines burn gasoline, also known as AVGAS (Aviation gasoline).

왕복엔진은 가솔린, 즉 AVGAS(항공용 가솔린)를 연소한다.

It is specially formulated for use in aircraft engines.

이 연료는 항공기 엔진에서 사용하기 위해 특별히 제조된다.

Combustion releases energy in the fuel, which is converted into the mechanical motion of the engine.

연료에서 연소가 에너지를 방출하며, 이는 엔진의 기계적 운동으로 변환된다.

AVGAS of any variety is primarily a hydrocarbon compound refined from crude oil by fractional distillation.

어떤 종류의 AVGAS도 기본적으로 원유에서 분별 증류로 정제된 탄화수소 화합물이다.

Aviation gasoline is different from the fuel refined for use in turbine-powered aircraft.

항공용 가솔린은 터빈엔진 항공기에서 사용되는 연료와 다르다.

AVGAS is very volatile and extremely flammable, with a low flash point.

AVGAS는 매우 휘발성이 강하고 극도로 가연성이며, 인화점이 낮다.

Turbine fuel is a kerosene-type fuel with a much higher flash point, so it is less flammable.

터빈 연료는 인화점이 훨씬 높은 등유 유형의 연료로, 따라서 덜 가연성이 있다.

(2) Turbine Engine Fuel

Aircraft with turbine engines use a type of fuel different from that of reciprocating aircraft engines.

터빈 엔진을 장착한 항공기는 왕복 엔진을 장착한 항공기와 다른 종류의 연료를 사용한다.

Commonly known as jet fuel, turbine engine fuel is designed for use in turbine engines and should never be mixed with aviation gasoline or introduced into the fuel system of a reciprocating aircraft engine fuel system.

일반적으로 제트 연료로 알려진 터빈 엔진 연료는 터빈 엔진에서 사용하도록 설계되었으며, 항공 가솔린과 혼합하거나 왕복 엔진의 연료 시스템에 도입해서는 안 된다.

2) Aircraft Fuel Systems

Each aircraft fuel system must store and deliver clean fuel to the engine(s) at a pressure and flow rate able to sustain operations regardless of the operating conditions of the aircraft.

각 항공기 연료 시스템은 항공기의 운용 조건에 관계없이 엔진에 깨끗한 연료를 저장하고 전달해야 하며, 연료는 적정 압력과 유량으로 공급되어야 한다.

(1) Small Single-Engine Aircraft

Small single-engine aircraft fuel systems vary depending on factors, such as tank location and method of metering fuel to the engine.

소형 단발 항공기의 연료 시스템은 탱크 위치와 엔진으로 연료를 계량하는 방식과 같은 요소에 따라 달라진다.

A high-wing aircraft fuel system can be designed differently from one on a low-wing aircraft.

고익 날개 항공기의 연료 시스템은 저익 날개 항공기와 다르게 설계될 수 있다.

An aircraft engine with a carburetor has a different fuel system than one with fuel injection.

기화기가 장착된 항공기 엔진은 연료 분사가 장착된 엔진과 다른 연료 시스템을 가지고 있다.

(2) Large Reciprocating-Engine Aircraft

Large, multiengine transport aircraft powered by reciprocating radial engines are no longer produced.

대형 다발 엔진 항공기는 왕복 성형(방사형) 엔진으로 구동되며 더 이상 생산되지 않는다.

However, many are still in operation.

하지만 많은 항공기가 여전히 운항 중이다.

They are mostly carbureted and share many features with the light aircraft systems previously discussed.

이 항공기들은 대부분 기화기를 사용하며, 이전에 논의한 경항공기 시스템과 많은 특성을 공유한다.

(3) Jet Transport Aircraft

Fuel systems on large transport category jet aircraft are complex with some features and components not found in reciprocating-engine aircraft fuel systems.

대형 수송기 제트 항공기의 연료 시스템은 복잡하며, 일부 특징과 구성 요소는 왕복 엔진 항공기의 연료 시스템에서는 찾을 수 없다.

They typically contain more redundancy and facilitate numerous options from which the crew can choose while managing the aircraft's fuel load.

이 시스템은 보통 더 많은 중복을 포함하며, 승무원이 항공기의 연료 부하를 관리하면서 선택할 수 있는 다양한 옵션을 제공한다.

Features like an onboard APU, single point pressure refueling, and fuel jettison systems, which are not needed on smaller aircraft, add to the complexity of an airliner fuel system.

소형 항공기에는 필요하지 않은 기내 APU, 단일 지점 압력 연료 급유, 연료 배출 시스템과 같은 특징들이 여객기 연료 시스템의 복잡성을 더한다.

Most transport category aircraft fuel systems are very much alike.

대부분의 수송기 범주 항공기 연료 시스템은 매우 유사하다.

Integral fuel tanks are the norm with much of each wing's structure sealed to enable its use as a fuel tank.

일체형 연료 탱크가 표준이며, 각 날개의 구조 대부분이 연료 탱크로 사용될 수 있도록 밀봉되어 있다.

Center wing section or fuselage tanks are also common.

중앙 날개 부분 또는 동체 탱크도 일반적이다.

These may be sealed structure or bladder type.

이들은 밀봉 구조일 수도 있고, 또는 방광형일 수도 있다.

Jet transport aircraft carry tens of thousands of pounds of fuel on board.

제트 수송기 항공기는 수만 파운드의 연료를 탑재한다.

Figure 4-22 shows a diagram of a Boeing 777 fuel tank configuration with tank capacities.

그림 4-22는 보잉 777 연료 탱크 배치도와 탱크 용량을 보여준다.

4-6. Aircraft Lighting Systems

1) Introduction

Aircraft lighting systems provide illumination for both exterior and interior use.

항공기 조명 시스템은 외부와 내부 모두를 위한 조명을 제공한다.

Lights on the exterior provide illumination for such operations as landing at night, inspection of icing conditions, and safety from midair collision.

외부 조명은 야간 착륙, 결빙 상태 점검, 그리고 공중 충돌 방지를 위한 작업에 조명을 제공한다.

Interior lighting provides illumination for instruments, cockpits, cabins, and other sections occupied by crewmembers and passengers.

내부 조명은 계기, 조종석, 객실, 그리고 승무원 및 승객이 있는 다른 구역에 조명을 제공한다.

Certain special lights, such as indicator and warning lights, indicate the operation status of equipment.

지시등과 경고등 같은 특정 특수 조명은 장비의 작동 상태를 나타낸다.

2) Exterior Lights

Position, anticollision, landing, and taxi lights are common examples of aircraft exterior lights.

위치(표시)등, 충돌방지등, 착륙, 그리고 유도등은 항공기 외부등의 일반적인 예이다.

Some lights are required for night operations.

일부 등은 야간 운항을 위해 필수적이다.

Other types of exterior lights, such as wing inspection lights, are of great benefit for specialized flying operations.

날개 점검등과 같은 다른 종류의 외부등은 특수 비행 작업에 큰 도움을 준다.

(1) Position Lights

Aircraft operating at night must be equipped with position lights that meet the minimum requirements specified by Title 14 of the Code of Federal Regulations.

야간에 운항하는 항공기는 연방항공규정(FAR) 제14조에서 명시된 최소 요구 사항을 충족하는 위치표시등을 장착해야 한다.

A set of position lights consists of one red, one green, and one white light.

위치표시등 세트는 빨간색, 초록색, 그리고 흰색 등으로 구성된다.

The green light unit is always mounted at the extreme tip of the right wing.

초록색 등은 항상 오른쪽 날개의 끝단에 장착된다.

The red unit is mounted in a similar position on the left wing.

빨간색 등은 왼쪽 날개의 비슷한 위치에 장착된다.

The white unit is usually located on the vertical stabilizer in a position where it is clearly visible through a wide angle from the rear of the aircraft.

흰색 등은 보통 수직 안정판에 위치하며, 항공기 후방에서 넓은 각도로 명확하게 보이는 위치에 있다.

(2) Anticollision Lights

An anticollision light system may consist of one or more lights.

충돌 방지등 시스템은 하나 이상의 등으로 구성될 수 있다.

They are rotating beam lights that are usually installed on top of the fuselage or tail in such a location that the light does not affect the vision of the crew member or detract from the visibility of the position lights.

이 등은 회전식 빔 등으로, 일반적으로 동체 상부 또는 꼬리에 설치되며, 이 위치는 빛이 승무원의 시야에 영향을 주거나 위치표시등의 가시성을 저해하지 않도록 되어 있다.

Large transport type aircraft use an anti-collision light on top and one on the bottom of the aircraft.

대형 수송기 유형의 항공기는 항공기 상부와 하부에 각각 충돌 방지등을 사용한다.

(3) Landing Lights

Landing lights are installed in aircraft to illuminate runways during night landings.

착륙등은 야간 착륙 시 활주로를 비추기 위해 항공기에 설치된다.

These lights are very powerful and are directed by a parabolic reflector at an angle providing a maximum range of illumination.

이 조명은 매우 강력하며, 포물선 반사체에 의해 최대 조명 범위를 제공하는 각도로 조정된다.

Landing lights of smaller aircraft are usually located midway in the leading edge of each wing or streamlined into the aircraft surface.

소형 항공기의 착륙등은 일반적으로 각 날개의 앞전 중간에 위치하거나 항공기 표면에 매끄럽게 통합된다.

Landing lights for larger transport category aircraft are usually located in the leading edge of the wing close to the fuselage.

대형 수송 카테고리 항공기의 착륙등은 일반적으로 동체 가까운 날개의 앞전에 위치한다.

(4) Taxi Lights

Taxi lights are designed to provide illumination on the ground while taxiing or towing the aircraft to or from a runway, taxi strip, or in the hangar area. [Figure 4-27]

유도등은 항공기를 활주로, 유도로, 또는 격납고 지역으로 이동하거나 견인할 때 지면을 밝히도록 설계되었다. [그림 4-27]

Taxi lights are not designed to provide the degree of illumination necessary for landing lights.

유도등은 착륙등이 필요한 수준의 조명을 제공하도록 설계되지 않았다.

On aircraft with tricycle landing gear, either single or multiple taxi lights are often mounted on the non-steerable part of the nose landing gear.

삼륜식 착륙 장치를 가진 항공기에서는 단일 또는 복수의 유도등이 조향되지 않는 전방 착륙 장치 부분에 장착되는 경우가 많다.

3) Interior Lights

Aircraft are equipped with interior lights to illuminate the cabin.[Figure 4-26]

항공기는 객실을 밝히기 위해 내부 조명을 갖추고 있다.[그림 4-26]

Often white and red light settings are provided.

종종 백색과 적색 조명 설정이 제공된다.

Commercial aircraft have a lighting system that illuminates the main cabin, an independent lighting system so that passengers can read when the cabin lights are off, and an emergency lighting system on the floor of the aircraft to aid passengers of the aircraft during an emergency.

상업용 항공기는 주 객실을 밝히는 조명 시스템, 객실 조명이 꺼져 있을 때 승객이 독서를 할 수 있도록 하는 독립 조명 시스템, 그리고 비상 상황 시 승객을 돕기 위한 항공기 바닥에 설치된 비상 조명 시스템을 갖추고 있다.

Chapter 5
Aircraft Engines

5-1. General

Aircraft engines can be classified by several methods.

항공기 엔진은 여러 가지 방법으로 분류될 수 있다.

They can be classed by operating cycles, cylinder arrangement, or the method of thrust production.

운전 사이클, 실린더 배열, 또는 추력 생성 방법에 따라 분류될 수 있다.

All are heat engines that convert fuel into heat energy that is converted to mechanical energy to produce thrust.

모든 엔진은 연료를 열 에너지로 변환하고 이를 기계 에너지로 변환하여 추력을 생성하는 열기관이다.

Most of the current aircraft engines are of the internal combustion type because the combustion process takes place inside the engine.

현재 대부분의 항공기 엔진은 연소 과정이 엔진 내부에서 이루어지기 때문에 내연기관이다.

Aircraft engines come in many different types, such as gas turbine based, reciprocating piston, rotary, two or four cycle, spark ignition, diesel, and air or water cooled.

항공기 엔진은 가스터빈 기반, 왕복 피스톤, 로터리, 2사이클 또는 4사이클, 스파크 점화, 디젤, 공냉 또는 수냉 방식 등 여러 유형이 있다.

Reciprocating and gas turbine engines also have subdivisions based on the type of cylinder arrangement (piston) and speed range (gas turbine).

왕복 엔진과 가스터빈 엔진은 실린더 배열(피스톤)과 속도 범위(가스터빈)에 따라 하위 분류가 있다.

5-2. Reciprocating Engines

The basic major components of a reciprocating engine are the crankcase, cylinders, pistons, connecting rods, valves, valve-operating mechanism, and crankshaft.

왕복엔진의 기본 주요 구성 요소는 크랭크케이스, 실린더, 피스톤, 커넥팅 로드, 밸브, 밸브 작동 메커니즘, 그리고 크랭크축이다.

In the head of each cylinder are the valves and spark plugs.

각 실린더의 헤드에는 밸브와 스파크 플러그가 있다.

One of the valves is in a passage leading from the induction system; the other is in a passage leading to the exhaust system.

밸브 중 하나는 흡입 시스템에서 이어지는 통로에 있고, 다른 하나는 배기 시스템으로 이어지는 통로에 있다.

Inside each cylinder is a movable piston connected to a crankshaft by a connecting rod.

각 실린더 내부에는 커넥팅 로드를 통해 크랭크축에 연결된 움직이는 피스톤이 있다.

5-3. Gas Turbine Engines

In a reciprocating engine, the functions of intake, compression, combustion, and exhaust all take place in the same combustion chamber.

왕복 엔진에서는 흡입, 압축, 연소, 배기가 모두 동일한 연소실에서 이루어진다.

Consequently, each must have exclusive occupancy of the chamber during its respective part of the combustion cycle.

따라서 연소 사이클의 각 단계 동안 각각이 연소실을 독점적으로 사용해야 한다.

A significant feature of the gas turbine engine is that separate sections are devoted to each function, and all functions are performed simultaneously without interruption.

가스터빈 엔진의 중요한 특징은 각 기능에 별도의 구역이 할당되어 있으며, 모든 기능이 중단 없이 동시에 수행된다는 것이다.

A typical gas turbine engine consists of air inlet, compressor section, combustion section, turbine section, exhaust section, accessory section.

일반적인 가스터빈 엔진은 공기 흡입구, 압축기 구역, 연소 구역, 터빈 구역, 배기 구역, 부속 구역으로 구성된다.

The systems necessary for starting, lubrication, fuel supply, and auxiliary purposes, such as anti-icing, cooling, and pressurization.

시동, 윤활, 연료 공급 및 제빙, 냉각, 압력 유지와 같은 보조 목적으로 필요한 시스템들이 포함된다.

Chapter 6
Propellers

6-1. General

The propeller, the unit that must absorb the power output of the engine, has passed through many stages of development.

프로펠러는 엔진의 출력 동력을 흡수해야 하는 장치로, 많은 발전 단계를 거쳐왔다.

Although most propellers are two-bladed, great increases in power output have resulted in the development of four-and six-bladed propellers of large diameters.

대부분의 프로펠러는 두 개의 브레이드를 가지고 있지만, 출력 동력의 큰 증가로 인해 큰 직경을 가진 4엽 및 6엽 프로펠러가 개발되었다.

There are several forces acting on the propeller as it turns; a major one is centrifugal force.

프로펠러가 회전할 때 여러 가지 힘이 작용하는데, 그 중 중요한 힘은 원심력이다.

This force at high rpm tends to pull the blades out of the hub, so blade weight is very important to the design of a propeller.

이 힘은 고속 회전 시 브레이드를 허브에서 끌어내는 경향이 있기 때문에, 브레이드의 무게는 프로펠러 설계에서 매우 중요하다.

As aircraft speeds increased, turbofan engines were used for higher speed aircraft.

항공기 속도가 증가함에 따라, 더 높은 속도의 항공기를 위해 터보팬 엔진이 사용되었다.

Propeller-driven aircraft have several advantages and are widely used for applications in turboprops and reciprocating engine installations.

프로펠러가 구동하는 항공기는 여러 가지 장점이 있으며, 터보프롭과 피스톤 엔진 설치에 널리 사용된다.

Takeoff and landing can be shorter and less expensive.

이착륙 거리가 짧고 비용이 덜 들 수 있다.

New blade materials and manufacturing techniques have increased the efficiency of propellers.

새로운 날개 재료와 제조 기술은 프로펠러의 효율성을 증가시켰다.

Many smaller aircraft will continue to use propellers well into the future.

많은 소형 항공기들은 앞으로도 계속해서 프로펠러를 사용할 것이다.

6-2. Type of Propellers

1) Pusher Propellers

Pusher propellers are those mounted on the downstream end of a drive shaft behind the supporting structure.

추진식 프로펠러는 구동 샤프트의 하류 끝에 지지 구조물 뒤에 장착된 프로펠러이다.

Pusher propellers are constructed as fixed- or variable-pitch propellers.

추진식 프로펠러는 고정 피치 또는 가변 피치 프로펠러로 제작된다.

Seaplanes and amphibious aircraft have used a greater percentage of pusher propellers than other kinds of aircraft.

수상 비행기와 수륙 양용 항공기는 다른 종류의 항공기보다 더 많은 비율로 푸셔 프로펠러를 사용해왔다.

2) Fixed-Pitch Propeller

Generally, this type of propeller is one piece and is constructed of wood or aluminum alloy.

일반적으로 이 유형의 프로펠러는 하나의 조각으로 되어 있으며, 나무 또는 알루미늄 합금으로 제작된다.

Fixed-pitch propellers are designed for best efficiency at one rotational and forward speed.

고정식 피치 프로펠러는 하나의 회전 속도와 전진 속도에서 최상의 효율을 낼 수 있도록 설계된다.

They are designed to fit a set of conditions of both airplane and engine speeds and any change in these conditions reduces the efficiency of both the propeller and the engine.

이 프로펠러는 항공기와 엔진의 속도 조건에 맞게 설계되며, 이러한 조건이 변경되면 프로펠러와 엔진의 효율이 모두 감소한다.

The fixed-pitch propeller is used on airplanes of low power, speed, range, or altitude.

고정식 피치 프로펠러는 낮은 출력, 속도, 비행 범위 또는 고도의 항공기에 사용된다.

Many single-engine aircraft use fixed-pitch propellers and the advantages to these are less expense and their simple operation.

많은 단발기 항공기들이 고정식 피치 프로펠러를 사용하며, 이 프로펠러의 장점은 적은 비용과 간단한 작동이다.

This type of propeller does not require any control inputs from the pilot in flight.

이 유형의 프로펠러는 비행 중 조종사로부터의 조작 입력이 필요 없다.

김형래

- 한국항공대학교(공학사)
- 국립경상대학교(공학석사)
- 국립경상대학교(공학박사)
- 대한민국 공군 정비장교 근무
- 한국항공우주산업(주) 근무
- 한국폴리텍대학 항공캠퍼스
  항공기계과 교수
- 현재, 한국폴리텍대학 남인천캠퍼스
  항공MRO과 교수

자격

- 항공기체기술사, 항공기사
- 항공정비사
- 기술지도사

저서

- 항공기술영어(회화편), 연경문화사, 2011
- 항공기 정비 영어, 연경문화사, 2024

미래 항공종사자들을 위한 항공기 정비 영어 입문서

항공기 정비 영어 기본편
Aircraft Maintenance English Basic

발행일	2025년 2월 19일
저자	김형래
펴낸이	이정수
책임 편집	최민서 · 신지항
펴낸곳	연경문화사
등록	1-995호
주소	서울시 강서구 양천로 551-24 한화비즈메트로 2차 807호
대표전화	02-332-3923
팩시밀리	02-332-3928
이메일	ykmedia@naver.com
값	18,000원
ISBN	978-89-8298-220-0 (93550)